Grasping the water, energy, and food security nexus in the local context

Case study: Karawang Regency, Indonesia

Aries Purwanto

Grasping the water, energy, and food security nexus in the local context

Case study: Karawang Regency, Indonesia

Thesis
submitted in fulfilment of the requirements of
the Academic Board of Wageningen University and
the Academic Board of the IHE Delft Institute for Water Education
for the degree of doctor
to be defended in public
on Tuesday, 15 June 2021 at 4:00 pm
in Delft, the Netherlands

by

Aries Purwanto

Born in Karawang, Indonesia

CRC Press/Balkema is an imprint of the Taylor & Francis Group, an informa business

© 2021, Aries Purwanto

Although all care is taken to ensure integrity and the quality of this publication and the information herein, no responsibility is assumed by the publishers, the author nor IHE Delft for any damage to the property or persons as a result of operation or use of this publication and/or the information contained herein.

A pdf version of this work will be made available as Open Access via https://ihedelftrepository.contentdm.oclc.org/. This version is licensed under the Creative Commons Attribution-Non Commercial 4.0 International License, http://creativecommons.org/licenses/by-nc/4.0/

Published by:
CRC Press/Balkema
Pub.NL@taylorandfrancis.com
www.crcpress.com – www.taylorandfrancis.com

ISBN: 978-1-032-07645-4 (Taylor & Francis Group)
ISBN: 978-94-6395-702-1 (Wageningen University)
DOI: https://doi.org/10.18174/541237

To my beloved father and my lovely wife and children

In memory of my beloved late mother

SUMMARY

Despite an abundance in natural and human resources, Indonesia has not been able to significantly improve the level of water, energy, and food (WEF) security. Challenges in achieving WEF security targets mostly relate to resource mismanagement, lack of coordination, authority imbalance among sectors, and overlapping roles and responsibility among levels of government. The immature process of the decentralization has made these challenges even more complex; it is not easy to unify the vision of local governments (i.e. 34 provinces, 416 regencies, 98 cities) with the Regional Head who is elected every 5 years and sometimes brings different goals and development approaches.

The lack of information, awareness, coordination and a common framework to bridge the gaps between national and local governments, jeopardizes the attainment of WEF security targets which have been set in the national long-term planning (RPJPN) and mid-term planning (RPJMN). Unfortunately, this complex issue has not received the attention it deserves, from a scientific perspective nor from a practical implementation point of view such as through laws, policies and planning processes.

This research addresses these knowledge and implementation gaps by analysing the interlinkages among variables in the WEF system using the nexus approach which integrates management and governance across sectors and scales. The main objective of this research is to grasp the WEF security nexus in the local context and to evaluate the implications of planned local interventions in WEF sectors by developing a conceptual and quantitative analysis framework and employing system dynamics modelling through a stakeholder engagement and co-development process. The Karawang Regency in Indonesia is chosen as an illustrative case study as it represents all challenges and variables at the local level of WEF security nexus.

The first part of this study identifies knowledge gaps and common critiques on the WEF nexus framework that have emerged since the concept was proposed. It analyses current improvements of the WEF nexus concept, applications and impacts during the period of 2012-2020. By reviewing 10 existing WEF nexus frameworks, several gaps and omissions as well as their possible improvements are identified. Four principles that must be of serious consideration in developing the future WEF nexus framework and improving the WEF nexus-related studies are proposed, i.e. to make them more understandable, to make them adaptable to many diverse situations, to ensure reliable and valid data, and to be applicable across scales. The perspective of "from local to global" and locally-based WEF resource management are also suggested to ensure that WEF security can be achieved sustainably in local communities and they will help towards national and global targets.

Secondly, several strategies and their practical implementation for WEF-related sectors in the study area are formulated using the composite method of Location Quotient (LQ) and Competitive Position (CP). This method assesses the agglomeration level and growth potential in each WEF sector, locating them in a four-quadrant matrix. Quadrant I, signifying the advantaged cluster, contains 1 sector (energy-related sector); quadrant II, signifying the potential cluster contains 1 sector (water-related sector) and 2 sub-sectors (estate & horticulture crops and water supply sub-sectors); none of sector in quadrant III or capable cluster, while 1 sector (food-related sector) and 4 sub-sectors (food crops, electricity, livestock, and fishery sub-sectors) fall in quadrant IV, the disadvantaged cluster. The analysis shows that the general characteristics of WEF-related sectors in a region can be clearly distinguished based on its main economic development focus. This preliminary economic-based evaluation gives a better understanding and more comprehensive insights for policy-makers and other stakeholders, although the clear interrelation among variables and sectors is not assessed at this stage of analysis.

The third set of results shows that participatory or group model building is beneficial in assisting local stakeholders to grasp the complexity of the WEF security system. The group model building approach covers all major internal and external factors and drivers, including possible feedback mechanisms and key variables to be further analysed. A qualitative Karawang WEF security (K-WEFS) model is established with stakeholders, and is composed of six sub-models with water, energy and food sectors as internal factors and population, economy and ecosystem services as external drivers. The collaborative action plan, using system dynamics analysis and group model building, not only can be implemented in WEF sectors but also other development planning and policy-making process such as infrastructure, trade and services, monetary, transportation etc.

Fourthly, building on the qualitative K-WEFS model, a quantitative stock-flow diagram (SFD) is developed. By employing STELLA® professional software, three planned policy interventions in WEF-related sectors are modelled in an integrated way. The impacts on the available resources per person (APP) and self-sufficiency levels (SSL) of resources are analysed in four scenarios, including business as usual and several combinations of planned interventions. Implications are based on model simulation, while possible practical actions are derived from both model simulation and other considerations, such as local planning ambitions, national programs, local experts and modeller's opinion. Several potentially unanticipated and indirect impacts of policy interventions are also highlighted in this quantitative simulation.

Results and findings in this study, derived from the K-WEFS nexus framework are expected to assist the local planner and decision-makers to deal with challenges in WEF resource management by making trade-offs explicit, building synergies among WEF-related sectors and eventually improving the WEF security target's achievement.

CONTENTS

Summary vii

Contents ix

1   Introduction 1

    1.1 Background 1

        1.1.1 WEF security nexus concepts 2

        1.1.2 WEF security in Indonesia 3

    1.2 Problem identification 3

    1.3 Research objectives 4

    1.4 Methods summary 5

    1.5 Study area 6

    1.6 Thesis outline 8

2   WEF nexus: Knowledge gaps, criticisms, and improvements 9

    2.1 Introduction 10

    2.2 Assessing existing WEF nexus frameworks 11

    2.3 Literature criticisms on the WEF nexus concept 17

    2.4 Efforts to narrow gaps and address criticisms 20

        2.4.1 Narrowing knowledge gaps 20

        2.4.2 Addressing criticisms in WEF nexus frameworks 22

    2.5 Updating the WEF nexus frameworks 26

    2.6 Conclusion 27

3   WEF-related sectors in local economic development 29

    3.1 Introduction 30

    3.2 Methods 31

        3.2.1 Dataset 31

        3.2.2 Location quotient (LQ) 32

        3.2.3 Competitive position (CP) 34

    3.3 Study area 35

    3.4 Result and discussion 38

        3.4.1 Comparison of sector's agglomeration 38

        3.4.2 Analysis of water, energy, and food-related sectors ... 42

        3.4.3 Trends of WEF-related sub-sectors in Karawang Regency ... 44

        3.4.4 Proposed strategies for WEF-related sectors and sub-sectors in Karawang Regency 47

        3.4.5 Sustainability and nexus perspective in WEF resource management .. 52

3.5   Conclusion.. 53

4   Group model building on qualitative WEF security nexus dynamics 55

4.1   Introduction 56

4.2   Methods 58

 4.2.1   Causal loop diagram (CLD)... 58

 4.2.2   Group model building.. 59

4.3   Results & discussion 61

 4.3.1   State of the system 61

 4.3.2   Basic concept of WEF security nexus in Karawang Regency.............. 61

 4.3.3   GMB workshop script... 62

 4.3.4   WEF security sub-models.. 66

 4.3.5   Karawang WEF security (K-WEFS) model 68

4.4   Conclusion... 75

5   Quantitative simulation of WEF security nexus 77

5.1   Introduction 78

5.2   Methods 80

 5.2.1   Stock Flow Diagrams (SFDs)... 80

 5.2.2   Study area 80

 5.2.3   Model development process 82

 5.2.4   Model validation and policy sensitivity simulation............................ 85

5.3   Results 89

 5.3.1   K-WEFS nexus model structure 89

 5.3.2   Model behaviour test results 91

 5.3.3   Quantitative K-WEFS model analyses under BAU conditions............ 92

 5.3.4   Sensitivity and policy scenario analysis 96

5.4   Discussion 101

5.5   Conclusion.. 106

6   Synthesis and recommendation 107

6.1   General conclusions 108

6.2   Synthesis... 109

 6.2.1   Increasing resource supply and efficiency........................... 109

 6.2.2   Reducing trade-offs 111

 6.2.3   Building institutional and intergovernmental synergies 114

6.3   Main contributions 116

6.4   Recommendations 119

 6.4.1   Recommendation for further scientific research................... 119

 6.4.2   Practical recommendation 120

Appendix 123

References 133

List of Acronyms... 149

List of Tables... 151

List of Figures 153

Acknowledgements 157

About the Author... 159

1

INTRODUCTION

1.1 BACKGROUND

The existence of water, energy and food are important for people to achieve welfare, alleviate poverty and reach a sustainable level of development (FAO, 2014). Based on world projection in the next decades, Hoff (2011) concludes there will be a prominent increase of water, energy, and food demand due to the pressure of population growth, economic activities, the changes in diets, culture, technology, and climate. In addition, notwithstanding the global development process has quickly sped up over the past 50 years, there is still lack of equity in distribution between and within countries in gaining advantages especially in water, energy and food security. Water, energy, and food security-related challenges are getting urgent and need to be resolved simultaneously in an integrative manner. This is exacerbated by diminishing resource bases both in quantity and quality due to natural and human activities pressuring these resources. De Fraiture & Wichelns (2010) underline that increasing water demand for urban, industrial and environmental protection will escalate competition with the rising need of water for agriculture. Globally, the energy demand will almost double, while the demand of water and food are foreseen to escalate by more than 50% in 2050 (IRENA, 2015).

The World Food Summit in 1996 defined food security as where 'all people at all times have physical and economic access to sufficient safe and nutritious food to meet their dietary needs and food preferences for an active and healthy life'. Food production, food accessibility and food quality are three key elements in food security. Almost in the same way, water security as described by UN-Water is 'the capacity of a population to safeguard sustainable access to adequate quantities of acceptable quality water for sustaining livelihoods, human well-being, and socio-economic development, for ensuring protection against water-borne pollution and water-related disasters, and for preserving ecosystems in a climate of peace and political stability" (UN-Water, 2013). In terms of energy security, the International Energy Agency (IEA) in 1974 defined it in a clear way as 'an uninterrupted availability of energy sources at an affordable price'. Hoff (2011) underlines that resources availability is not the one and only factor of security.

Accessibility and quality are also crucial to be covered especially in the extreme condition that exist naturally, economically, and socially. Resource availability can be defined as physical existence of the resource to meet demand in all level (from household to national level). Furthermore, accessibility of resource means that the resource is easily to obtain and in affordable price, while quality aspect interpreted as the ability of the resource to meet quality standard which has been set for certain purpose such as the guideline on drinking water quality established by World Health Organization (see WHO, 2017).

Water, energy, and food security are becoming a major topic that is vigorously discussed not only in developing countries but also in developed countries. The interaction among their components internally and interconnection with environment condition, social, governance and even political situation make this issue is immensely complex. Achieving certain levels of water, energy and food security simultaneously is a complex challenge that will influence, and is influenced by other sectors including social, political, and environmental condition (Bizikova *et al.*, 2013; Endo *et al.*, 2015). Resolving one problem partially without considering its interlinkage will only shift problems from one resource perspective to another and may cause unexpected effects (Kenway *et al.*, 2011; Bizikova *et al.*, 2013; FAO, 2014; El Gafy *et al.*, 2016). Additionally, focusing only on one certain aspect of security, without considering others may also cause unbalanced supply and ineffective target achievement, or could even damage the sustainable us of other resources.

1.1.1 WEF security nexus concepts

The basic concept of the water, energy, and food security nexus approach has been developed and extensively discussed in Bonn 2011 Conference. In its background paper, the nexus approach is defined as 'an approach that integrates management and governance across sectors and scales' (Hoff, 2011). The paper also provided evidence that this approach is effective to enhance water, energy, and security by improving efficiency, lowering trade-offs, developing synergies, and improving governance. Nevertheless, there are still knowledge gaps in this approach including analytical framework disharmony for overcoming institutional disjunctions and power imparity among sectors. There is no sole technique able to be applied for every specific circumstance suitably (Endo *et al.*, 2015). Thus, to deal with different and specific situation in each region, deconstruction of the nexus approach (Lele *et al.*, 2013) and specific context elaboration (El Gafy *et al*, 2016) have to be considered in order to make more effective and contextualized solutions on water, energy, and food security and to assist decision makers in managing resources. The WEF security nexus framework established by Holger Hoff becomes the main reference of many WEF nexus studies all over the world.

1.1.2 WEF security in Indonesia

Indonesia has almost all the resources that are needed in achieving WEF security such as oil, coal, natural gas, abundant solar radiation, water resources, and also land resources to produce foods. In Indonesia, considerable attention has been paid to the level of WEF security with several international publications documenting the current condition. The Asian Development Bank (ADB), for instance, ranked the National Water Security Index of Indonesia as 27 out of 48 Asian Countries (ADB, 2016a). An unfavourable position also obtained regarding energy security, where Indonesia ranked 85 of 125 countries in Energy Trilemma Index (World Energy Council, 2016). Indonesian rank in food security was in 71 out of 113 countries by 2016, left behind other Southeast Asian countries like Singapore, Malaysia, Thailand and Vietnam (The Economist Intelligence Unit, 2016) while in Global Hunger Index 2016, Indonesia categorized as 'serious' with score 21.9 (IFPRI, 2016). There is "a silent but imminent crisis in food, as well as in energy, and water supplies" says Rahmadi (2013). Symptoms concerning water, energy and food insecurity are increasingly visible. Indonesia is one of agricultural product (grains, horticultures, and livestock) importers, while those commodities are potentially produced in this country. Floods in the wet season and droughts in the dry season are regularly happened. Furthermore, water pollution due to domestics and industrial activities, agricultural land conversion and climate change are even worsen the WEF insecurity. Resources exploitation without balanced and proper management which considers other related factors and sectors is the big reason and may cause resource insecurity in Indonesia. This is related to the problem identification in the next section.

1.2 PROBLEM IDENTIFICATION

Water, energy, and food security status of Indonesia are in a disadvantaged situation, whereas Indonesia has abundant potential resources to support those securities. Principally, the Indonesian government has strong commitments to achieve its targets in water, energy, and food security as outlined in national long-term and medium-term planning. However, several acute problems have obstructed the attainment of its national targets, i.e. resources mismanagement, lack of coordination, and authority imbalance among sectors, levels and scales (Bellfield *et al.*, 2016).

The challenges are even getting more complex in the current decentralization era where local governments (i.e. 34 provinces, 416 regencies, and 98 cities) have also their specific local approaches and targets. Decentralized systems can be an effective way to achieve national targets if each local government has a harmonious framework and perspective on it. Otherwise, the gaps among sectors and levels are getting wider, and the targets are becoming more difficult to reach. Quincieu (2015) emphasizes the need of preferable and clearer roles, responsibilities, programmes and policies among district, provincial and central government in Indonesia. An integrated local framework in evaluation and planning is importantly needed to unify visions of each local government in achieving both local and national water, energy, and food security targets (Figure 1.1).

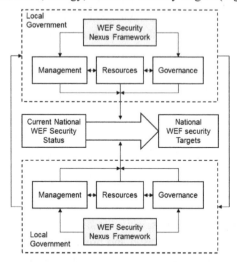

Figure 1.1. The importance of WEF security nexus framework

Unfortunately, studies in water, energy, and food security nexus have been carried out mostly in the perspectives of global and national level. It has not sufficiently addressed to the local government viewpoint especially in a developing country like Indonesia. This research attempts to fill one of those gaps, by focusing to analyse, develop a framework and model, and assess the water, energy, and food security nexus in a local perspective in Indonesia.

1.3 RESEARCH OBJECTIVES

The major objective of this research is to grasp the WEF security nexus in the local context and evaluate the implications of planned local scale interventions in WEF sectors by developing a conceptual and quantitative analysis framework together with local stakeholders. Furthermore, the specific objectives of the research have been formulated to:

1) analyse the existing WEF security nexus concepts and frameworks (Chapter 2);

2) assess the WEF-related sectors and their behaviour in the local economic development (Chapter 3);

3) develop a qualitative causal loop mapping of WEF security nexus by involving local WEF-related stakeholder through group model building (Chapter 4);

4) develop a quantitative system dynamics model WEF security nexus in local context (Chapter 5);

5) evaluate the local development policies and planning on the water, energy and food related sectors (Chapter 5 & 6).

All the objectives have addressed the central question in this study that asks the implications of planned interventions in Karawang Regency to local water, energy, and food security, and its contribution to the achievement of national targets.

1.4 METHODS SUMMARY

In general, this research will be organized using a mixed method that combines aspects relying on qualitative and quantitative approaches. The following Table 1.1 summarizes the steps and methods that have been employed in this study.

Table 1.1. Summary of methodology

No.	Stage	Method/Source
1.	Literature review	
	a. WEF security concepts	Hoff (2011); (FAO, 2014), etc.
	b. System dynamics	Forrester (1961); Sterman (2000), etc.
2.	Data collection	Primary and secondary data (BPS/Statistics Agency; ADB; etc.)
3.	Economic base analysis	▪ Analysing GRDP using Location Quotient (LQ) (Wang and Hofe, 2007) $$SLQ = \frac{X_{in}/Y_n}{X_i/Y}$$ ▪ Agglomeration growth (P) $$P = \left(\frac{LQ_{itn} - LQ_{it0}}{LQ_{it0}}\right) x 100\%$$ ▪ Competitive Cluster Bubble Chart (Pearce and Robinson, 2001; Zhao et al., 2016)
4.	Model conceptualization	Endogenous and exogenous indicators determination based on result of literature review and WEF security nexus diagram
5.	CLD development (Qualitative)	Sub-CLDs building and integration of each Sub-CLDs into integrated CLD WEF security by applying Group Model Building (Vennix, 1996)
6.	SFD development (Quantitative)	▪ Stock, flow, and variable determination ▪ Providing rates of change ▪ Governing equations
7.	Model validation	▪ Model behaviour test, sensitivity analysis and policy sensitivity with case study in Karawang Regency (Base year 2010-2019)

No.	Stage	Method/Source
		▪ Expert/stakeholder opinion

$$M=\frac{\sum(X_m-X_d)}{\sum X_d} \qquad MAE=\frac{1}{n}\sum|X_m-X_d|$$

$$R^2=\left(\frac{COv(X_m-X_d)}{\sigma X_m - \sigma X_d}\right)^2 \qquad MAPE$$

$$U_0=\frac{\sqrt{\sum(X_m-X_d)^2}}{\sqrt{\sum X^2_m}+\sqrt{\sum X^2_d}} \qquad =\frac{1}{n}\sum\left|\frac{X_m-X_d}{X_d}\right|$$

$$RMSE=\sqrt{\frac{1}{n}\sum(X_m-X_d)^2}$$

No.	Stage	Method/Source
8.	Planned intervention, policy analysis, and framework accomplishment	Using developed system dynamics model to analyse water conservation (artificial ponds), solar electricity development, agricultural land conversion, and other interventions.

1.5 STUDY AREA

Karawang Regency, Indonesia (Figure 1.2) was chosen as the case study in this research. This regency is one of the largest agricultural centres in Indonesia with Paddy as the main crop. On the other hand, this region also focuses on industrial development as stated in its long-term (twenty-yearly) planning year 2005-2025. Its vision is to achieve a prosperous region based on agricultural and industrial development. Secondly, this regency comprises urban, rural, and peri-urban area that have different characteristics to be managed specifically and comprehensively. Thirdly, the location is relatively close to Jakarta, the capital of Indonesia. It has become another centre of development activities besides industry (e.g. residential, infrastructures, trade and services, hotels, and other supporting facilities) that potentially increase the interaction between water, energy and food sectors, as well as the demand on those resources.

Karawang comprises paddy field of 95,287 Ha (54.4%), non-paddy agriculture of 38,805 Ha (22.1%), other land uses such as roads, houses, industries, and water bodies of 41,167 Ha (23.5%) (BPS of Karawang, 2019). The main commodities of horticulture products are cucumber, beans and mushroom, whereas for livestock, cows, buffaloes, sheep, goats and chickens are raised in this area. There are also another food sources such as capture fisheries (8,871 tons), aquaculture (44,024 tons), and salt production (3,981 tons) (BPS of Karawang, 2019). By 2016, around 954 units of large manufacturing companies, and approximately 9290 units of intermediate and small firms already existed in Karawang Regency. Furthermore, the total area provided by the government of Karawang to develop industrial estates is become one of the largest among other regions in South East Asian countries. It indicates that industrialization policy is a major policy direction both locally and nationally.

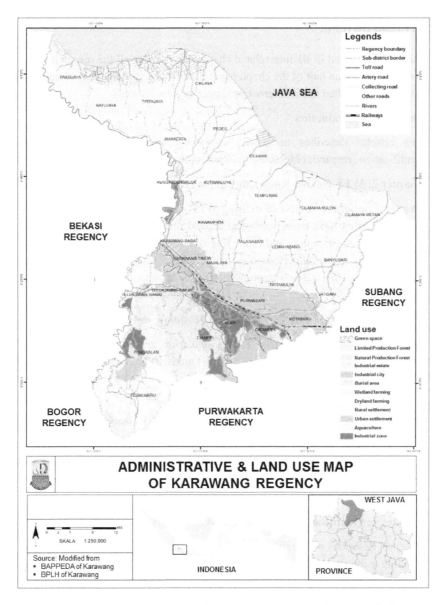

Figure 1.2. Administrative and land use map of Karawang Regency, Indonesia

This region is expected to represent the complexity of water, energy, and food interactions, so that all variables needed in developing system dynamics model both qualitative and quantitative can be derived comprehensively.

1.6 THESIS OUTLINE

This thesis is organized in six interrelated chapters to address all the research objectives in this study. More than half of the chapters are based on or adapted from research papers that have been published in the international peer-review journals.

1) **Chapter 1: Introduction**

   This chapter describes in detail the background of the research, problem identification, research objectives, general methods, study area, and thesis structure.

2) **Chapter 2: WEF Nexus: Knowledge gaps, criticisms, and improvements**

   Chapter 2 presents a review of existing WEF security nexus concepts, models and frameworks that have been developed around the world, including the knowledge gaps, criticisms, innovation and future improvements on this concept. The aim of this part is to bring a general perspective on how WEF resources should be managed and how the future WEF nexus framework, research, and implementation should be developed.

3) **Chapter 3: WEF-related sectors in local economic development**

   This chapter consist of the explanation, calculation, and analysis on economic base sector using composite methods of Location Quotient (LQ) and Competitive Position (CP) to evaluate the agglomeration level and growth and to determine some possible sustainable strategies in water, energy and food related sectors.

4) **Chapter 4: Group model building on qualitative WEF security nexus dynamics**

   This chapter elaborates the process a qualitative causal loop model development of a water, energy, and food (WEF) security nexus system to be used in analysing the interlinkages among WEF and other sectors by engaging all related local stakeholders through a group model building (GMB).

5) **Chapter 5: Quantitative simulation of WEF security nexus**

   This part describes the analysis of the water, energy, and food-related policies and planning using the developed quantitative framework (K-WEFS model) and also the evaluation of local policies and planning on WEF sectors.

6) **Chapter 6: Synthesis and recommendation**

   This chapter concludes the final findings of the research and provides main contributions, further research, and practical recommendation for the improvement of WEF security management, evaluation, and planning in local context. It also includes institutional and governance coherence in all levels to support national WEF security targets.

2

WEF NEXUS: KNOWLEDGE GAPS, CRITICISMS, AND IMPROVEMENTS

Abstract

This chapter presents knowledge gaps and critiques on the water-energy-food (WEF) nexus that have emerged since the concept of the WEF nexus was proposed by the World Economic Forum and the Bonn 2011 Conference. Furthermore, this study analyses current innovations on the WEF nexus concept, applications, and impacts during the period of 2012-2020. This begins by reviewing ten WEF nexus frameworks developed by international organizations and researchers. Based on this, several gaps and omissions in nexus frameworks are obvious in almost all developed frameworks. Studies that start to address some of these gaps are analysed, but are relatively few, and do not address all gaps. Several proposed improvements to nexus frameworks are identified to narrow the gaps and put the concept into practical implementation in WEF resources management and governance. Four principles and the perspective of "from local to global" for future WEF nexus framework development and analysis are suggested to ensure that the security of water, energy, and food resources can be achieved sustainably in local communities. This will aid the impact towards national and global ambitions on WEF security.

Keywords: water-energy-food (WEF), nexus concept, knowledge gaps, critiques, WEF security

This chapter is based on:

Purwanto A., Sušnik J., Suryadi F.X., de Fraiture C., (2021), *Water-energy-food nexus: critical review, practical applications, and prospects for future research*. Sustainability, 13, 1919, MDPI, DOI: https://doi.org/10.3390/su13041919 (Published).

2.1 INTRODUCTION

The connections between the water, energy, and food (WEF) sectors, known as the WEF nexus, are becoming a major academic, policy and societal topic that is increasingly discussed in global society, including the relationship with ecosystems, livelihoods, and the economy (e.g. de Fraiture *et al.*, 2010; Sušnik, 2018; Hülsmann *et al.* 2019). The challenges to manage water, energy and food resources simultaneously and meet multiple potentially conflicting objectives, without compromising the resource base of any sector are urgent and need to be resolved as best as possible (i.e. causing the least amount of damage to other sectors). This challenge demands an integrated approach in which the systems are considered as a whole. To add to the complexity, the WEF nexus influences and is influenced by other sectors including economic, social, political, and environmental conditions (Bizikova *et al.*, 2013; Endo *et al.*, 2015). The basic concept of the WEF security nexus approach was developed and extensively discussed in the Bonn 2011 Nexus Conference. In the resulting background paper, the nexus approach is defined as 'an approach that integrates management and governance across sectors and scales' (Hoff, 2011). The integration of theoretical approaches and practical implementation to solving policy challenges is urgently needed.

In Hoff's background paper (Hoff, 2011), initial guidance on how a nexus approach can improve the security of WEF resources by increasing efficiency, reducing trade-offs, building synergies and improving governance across sectors, including several policy recommendations was introduced. Since then, comprehensive studies and critical reviews by Endo *et al.*, (2017) and Albrecht *et al.*, (2018) state that the background paper by Holger Hoff and the World Economic Forum meeting in 2011 (WEF, 2011) has brought the topic and concept of the WEF nexus to the centre of global attention. However, several gaps were identified in nexus approach frameworks, and in subsequent nexus studies.

This chapter has several objectives. The first is to review existing WEF nexus frameworks, showing where they overlap and to consider key areas omitted from most, if not all of these frameworks. The second step is to revisit several key WEF nexus critiques, with the aim of identifying knowledge, research, and application gaps from those studies. These gaps are then mapped onto those from the nexus frameworks. These steps allow identification of the most urgent outstanding issues in current nexus research. Following this, several applied WEF nexus case studies are discussed, illustrating the extent to which the identified gaps and omissions have been addressed (or started to be addressed). From this, research gaps that are still present in nexus research are highlighted as urgent avenues for future research. The steps applied in this chapter are outlined in Figure 2.1. These steps are reflected in the structure of this chapter.

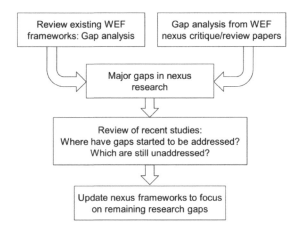

Figure 2.1. Schematic flow chart of the process followed in this chapter

2.2 ASSESSING EXISTING WEF NEXUS FRAMEWORKS

Existing WEF nexus frameworks were selected following a literature review of frameworks published in academic and non-academic sources using Google Scholar (cf. Walters, 2017), Science Direct, and Scopus databases. The databases were used to identify peer-reviewed scientific documents and other publications that employed the WEF nexus concept during the period 2012-2020. Further investigation was conducted to clarify the main concerns, key principles, and variables of the frameworks. While not meant as comprehensive, the results are representative of commonly presented WEF nexus frameworks in the literature, and much overlap can be identified between the frameworks discussed in this section.

While the concept of interlinkages and integration between the WEF sectors is not new (it has been understood by local communities, and the private sectors for some time e.g. Benson *et al.*, 2015; Bell *et al.*, 2016; Wichelns, 2017; Endo *et al.*, 2017), it is arguable that the modern concept of the WEF nexus became mainstreamed after 2011 when the World Economic Forum published a report which was the result of numerous analyses and studies (WEF, 2011). In the Global Risks 2011 (sixth edition) report, the interconnectedness between water, energy and food sectors with other external variables such as economic and population growth, environmental pressures, global governance failures, and even geopolitics conflict was postulated. This report identified some direct and indirect impacts that may arise due to risks associated with these interlinkages including major trends and uncertainties, levers and trade-offs (WEF, 2011). Several key were identified. These include recognizing trade-offs, integrated and multi-stakeholder planning, community level empowerment, market-led pricing, and technological and financial innovations to improve WEF management at any level.

After the publication of the 2011 Global Risks report, the next influential event was the Bonn 2011 Conference on the water, energy and food security nexus that was held on November 16-18, 2011. The background paper from the conference, entitled "Understanding the Nexus", has become an influential reference in research related to the WEF nexus approach (Hoff, 2011). Figure 2.2 shows the WEF nexus framework, describing the complexity of the WEF nexus with water availability as the core of the system. It considers the importance of sustainable development actions, global trends, and governmental interventions. Several key principles among others include resource productivity, the concept of waste as a resource, economic incentives, and coherence in governance, institutions, and policies.

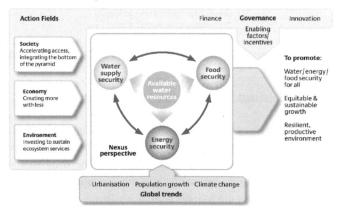

Figure 2.2. The water, energy, and food security nexus framework (source: Hoff, 2011. Reprinted with permission)

The Hoff background paper proposed some knowledge gaps in the nexus approach that were suggested to be addressed (Table 2.4). As the background paper is well known in the nexus research community, Hoff's knowledge gaps can become the focus of researchers in order to address the gaps and improve nexus understanding.

Many WEF frameworks, tools and models have been developed since the Bonn 2011 Conference. Several innovations and modifications of the 'original' WEF nexus framework have been developed by various international organizations, research institutes, and researchers. The association between WEF sectors and external variables, which is in line with the sustainable development concept, has been illustrated in various frameworks, some of which are summarized in Table 2.1.

Table 2.1. A selection of representative WEF nexus frameworks and their main features

WEF nexus framework	Document source & publisher
A Food, water, and energy nexus and the contribution of Himalayan ecosystem services	
■ The core of the framework is ecosystem goods and services to support WEF sectors, and implemented in South Asia ■ Key principles: (1) water storage capacity restoration, (2) climate and environmentally and social-friendly infrastructure development, (3) adequate investment for management, (4) incentive mechanism in managing ecosystem	■ Contribution of Himalayan ecosystems to water, energy and food security in South Asia: a nexus approach (Fig. 2, page 4, in this document source) ■ The International Centre for Integrated Mountain Development (ICIMOD, 2012)
B The Water-Energy-Land (WEL) nexus	
■ This framework widens the perspective of the nexus by considering land use competition for agriculture, forests, human settlement and infrastructure, and biodiversity, including the competition in water demands ■ Key principles: (1) rethinking the natural resources approach, (2) transformative action in addressing the demand, supply, efficiency, and resilience of natural resource, (3) integrated solution for an appropriate management of WEL	■ Confronting scarcity: Managing water, energy and land for inclusive and sustainable growth (Fig. 2.2, page 27, in this document source) ■ European Union (European Report on Development, 2012)
C The Resource Nexus	
■ This resource nexus framework focuses on five essential resources: water, energy, minerals, food, and land. ■ Key principles: (1) doubling resource efficiency, (2) transition toward sustainable energy systems (3) coordinating efforts to properly price resources, (4) rethinking of "the good life" and economic growth based on ever-increasing resource consumption, (5) working together to resolve disputes, (6) reinvesting in global leadership	■ The global resource nexus: The struggles for land, energy, food, water, and minerals (Fig. 1, page 7, in this document source) ■ The Transatlantic Academy (Andrews-Speed *et al.*, 2012)
D The CLEWS framework	
■ The framework integrates the assessment of three sectors of land, energy, and water resources using several tools i.e. LEAP by SEI, WEAP by SEI, and AEZ by IIASA and FAO models with climate change scenarios. ■ Key principles: (1) points identification at which the resource systems interact (2)	■ Integrated analysis of climate change, land-use, energy and water strategies (Fig. 1, page 622, in this document source) ■ Macmillan Publishers

WEF nexus framework	Document source & publisher
establish appropriate data exchanges between the modules, (3) Process repetition through a series of iterations	(www.nature.com/natureclimatechange); KTH-Royal Institute of Technology (Howells *et al.*, 2013)
E Nexus dialogue: agreed key interlinkages ■ The centre of the framework is ecosystems and climate and environment as external factors ■ Key principles: (1) Policy solutions (2) Land use management, (3) Cooperation agreements (4) Technology, operation and infrastructure, (5) Coordination and communication (6) Economic instruments (market-based or regulatory)	■ Reconciling resource uses in transboundary basins: assessment of the water-food-energy-ecosystems nexus (Fig. 5, page 22, in this document source) ■ The United Nations Economic Commission for Europe (UNECE, 2015)
F The framework linking water, food, and energy security ■ Ecosystem management as the core of the framework ■ Recommended policies: (1) integrated approach to policy design, (2) land and agricultural investment, (3) adaptive management of opportunities and risks ■ Stages: (1) assessing WEF security system, (2) envisioning future landscape, (3) investing in a WEF security, (4) transforming the system	■ The Water–Energy–Food Security Nexus: Towards a practical planning and decision-support framework for landscape investment and risk management (Fig. 6, page 15, in this document source) ■ The International Institute for Sustainable Development/IISD (Bizikova *et al.*, 2013)
G Approach to the Water-Energy-Food Nexus ■ This framework describes the complex interaction between human activities and natural resources with four main components; (1) Goals & interests, (2) resource base, (3) managing the nexus, and (4) drivers e.g. population, governance, climate change etc. ■ Key principles: (1) provide a stepwise process to address policy-making and intervention in a nexus way, (2) combine quantitative and qualitative assessment methods, (3) proposed indicators are based on available datasets (4) link intervention assessment to context status	■ Walking the nexus talk: Assessing the Water-Energy-Food Nexus in the Context of the Sustainable Energy for All Initiative (Fig. 1, page 13, in this document source) ■ The Food & Agriculture Organization/FAO (Flammini *et al.*, 2014)
H Key interactions between water, energy, food security ■ This framework identifies interaction between WEF security components within the existing national development planning with water security as the core system supported by forest as the main concern	■ How can Indonesia achieve security without eroding water, energy and food its natural capital? (Fig. 4, page 15, in this document source)

WEF nexus framework	Document source & publisher
▪ Key principles: (1) trade-offs between agricultural production and biofuel crops and deforestation (2) forest restoration and watershed protection, (3) interaction among four targets of WEF and forest in national development planning	▪ WCS Indonesia in partnership with the Global Canopy Programme (Bellfield *et al.*, 2016)
I Water, Land, Energy, Food & Climate (WLEFC) nexus	
▪ A systematic framework of scientific investigation, design of coherent policy goals, and instruments to deal with synergies, conflicts and related trade-offs from the interactions between WLEFC at bio-physical, socio-economic, and governance level. ▪ Key principles: (1) Policy coherence, (2) resource efficiency, (3) cross-sectoral governance, (4) interdisciplinary knowledge generation, (5) equal weight of each sector	▪ D1.5: Framework for the assessment of the nexus (Fig. 5, page 27, in this document source) ▪ Sustainable Integrated Management for the nexus of WLEFC for a resource-efficient Europe (SIM4NEXUS) (Ramos *et al.*, 2020)
J Main linkages within the land, water, and energy nexus	
▪ The framework indicates how the biophysical resources are interrelated to economic activities and a number of key policy objectives. It also considers the influence of socio-economic, climate change, and policies to the trade-offs and synergies in LWE nexus ▪ Key principles: 1st domain (LWE resources) represent biophysical system in term of quantity and quality, 2nd domain resources (goods and services) that meet the needs of the population e.g. agriculture, energy transformation, & water supply, 3rd domain highlight the resources nexus	▪ The land-water-energy nexus: Biophysical and economic consequences (Fig. 1.1, page 22, in this document source) ▪ The Organization for Economic Co-operation & Development (OECD, 2017)

The frameworks (Table 2.1) have been used as a reference at many levels of governance (e.g. global, national, regional, etc.) and spatial scales (e.g. basin scale, household scale, etc.) of management and planning. Because of their generic nature, many cannot be applied directly, with local-level modifications needed to capture specific circumstances. There is a need to make such comprehensive elaboration and adjustment by following the principles of context specific and stakeholder engagement to address the challenges and to make them more applicable to assist local level policymakers and other stakeholders.

Almost all the frameworks indicate that external factors need to be considered in managing WEF resources in an integrated manner. These can include climate change, population, and socio-economic development. The main differences between the frameworks include the key principles with regard to the main concern of each organization (e.g. water resources, food production, and stakeholder dialogue), the scales of each framework, and exogenous factors that influence and are deemed to be influenced by the WEF nexus. Some frameworks propose economic, social, and environmental issues as major components to be considered in managing resources. WCS Indonesia in partnership with the Global Canopy Programme (2016) for instance, focuses on forests as their main target. Those frameworks have followed context specific principles are arguably going against a true nexus framework where no resource should take 'centre stage'. Some frameworks do not comprehensively capture all interactions between variables in the WEF nexus due to: availability and limited access of data (Shannak *et al.*, 2018), inadequate consideration of politics in WEF resources management (de Grenade et al., 2016), under-representing gender perspectives (FAO, 2018), and ignoring livelihoods and development (Biggs et al., 2015).

The main issues covered by the frameworks in Table 2.1. are summarized in Table 2.2. Ten WEF nexus-based frameworks have been analysed and compared with 16 suggested features for the nexus concept. We have found that three out of 10 frameworks have good coverage (over half) of the 16 suggested features. The remaining seven cover up to half of the suggested features, implying they are less comprehensive or are more focused in their coverage.

Table 2.2. The evaluation of selected WEF nexus frameworks

No.	Main features in the nexus approach/ framework	WEF nexus frameworks (Table 2.1.)									
		A	B	C	D	E	F	G	H	I	J
1.	Incorporate WEF & exogenous variables, multi-resource (1,2,6)	V	V	V	V	V	V	V	V	V	V
2.	Social, economic and political context (3,5,6,9)		V	V	V	V	V	V	V	V	V
3.	Green economy, sustainability, environmental context (1,3)	V			V	V	V	V	V	V	V
4.	Interdisciplinary and transdisciplinary (4,6,8,9)	V	V		V	V	V	V		V	V
5.	Decision-making, policy-making, governance, solution-oriented (2,4,5,7)	V	V	V	V		V	V		V	V
6.	Incorporate global trends (1)			V	V	V	V	V	V		V	V
7.	Case study, local coverage, context specific, in-site context (1,2,3,6)	V			V	V		V	V	V	
8.	Capacity building, awareness raising (1,2)					V	V	V	V	V	
9.	Spatial-temporal scope (2,6)	V			V			V		V	
10.	Practical guide for implementation and simulation (5,7)				V	V		V		V	

No.	Main features in the nexus approach/ framework	WEF nexus frameworks (Table 2.1.)									
		A	B	C	D	E	F	G	H	I	J
11.	Mixed-methods of qualitative-quantitative (6,8)				V		V	V		V	
12.	Collaborative, participatory approach, stakeholders involvement (3,5,9)					V	V	V		V	
13.	Robust data sets, minimized data requirement (2,3,6)	V			V			V		V	
14.	Promote innovation, knowledge mobilization, theoretical approach (3,5,9)				V			V		V	
15.	Focuses on WEF resource security (1)	V						V	V		
16.	Appropriate & validated stages, using system approach and critical analysis (2,9)				V			V		V	

Notes: [1](Hoff, 2011), [2](Bazilian et al., 2011), [3](Keairns et al., 2016), [4](Endo et al., 2017), [5](Albrecht et al., 2018), [6](Shannak et al., 2018), [7](Dai et al., 2018), [8](Endo et al., 2019) [9](Urbinatti et al., 2020)

Based on Table 2.2, most existing frameworks include overlapping features. The efforts to bring the nexus concept into the process of policy and decision-making can also be seen in some frameworks.

Several features are distinctly lacking. These include:

(1) a focus on WEF resource security (i.e. availability, accessibility, quality of resources);

(2) appropriate & validated stages (cf. Bazilian *et al.*, 2011) of the WEF nexus modelling process, using a systems approach and critical analysis to better understand nexus complexity (Urbinatti *et al.*, 2020);

(3) promotion of innovation and knowledge mobilization;

(4) utilization of robust datasets from multiple sources; and

(5) participatory stakeholder involvement in framework development.

2.3 LITERATURE CRITICISMS ON THE WEF NEXUS CONCEPT

The WEF nexus concept has received criticism. The criticisms largely centre on the apparent lack of focus in nexus studies, the argument that the approach is not 'new' per-se, the lack of integration of some sectors (e.g. ecosystems), and the lack of common approaches to studying nexus problems. Regarding the concept, many researchers argue that the nexus is still an expanding concept (Smajgl et al., 2016), is relatively immature (Wichelns, 2017), that it is narrative but not useful in applications (Middleton et al., 2015) and that it is without any common definitions, methods and frameworks (Allouche et al., 2014; Endo et al., 2015; Benson et al., 2015; Albrecht et al., 2018). Furthermore, the application of the nexus concept has received much attention from various scholars. For example, Wichelns (2017) and Mitchell et al. (2015) warned that policy-making processes by applying the nexus approach and involving many stakeholders, especially in developing countries may lead to delays, slowness, and inertia. However, the involvement of stakeholders is deemed essential for proper nexus mapping and understanding. In addition, existing nexus implementations failed to address complex interlinkages due to lack of boundary definition (Galaitsi et al., 2018) and lack of data sharing and availability (Shannak et al., 2018). Critiques have been raised against the expected outcome in applying the nexus concept in various places around the world. Among the criticisms are the inability to consider inherent political factors (de Grenade *et al.*, 2016), the main democratic goal of sustainability (Biggs *et al.*, 2015), gender aspects and integration of programs, policies, and institutions at the national level (FAO, 2018), as well as the operationalization of WEF nexus in the decision-making process (Simpson and Jewitt, 2019). These issues are summarised in Table 2.3.

Table 2.3. Critiques on the concept, application, and implication of the WEF nexus

Main criticism	Reference
A. The WEF nexus concept	
1. It is not really a new [a] and/or novel [b] particularly in low level application or users such as farmers, fishers, etc. [c][d]	[a] Benson *et al.* (2015), [b] Bell *et al.* (2016), [c] Wichelns (2017), [d] Endo *et al.* (2017)
2. Sometimes seen as a 'nirvana' and narrative concept [a], 'mercurial' concept which led to an unpredictable changes in issue based on the context, location, and scale [b] without a means to address the challenges	[a] Middleton *et al.* (2015), [b] Bell *et al.* (2016
3. Only 'reframing resource scarcity as an existential threat' and 'cling to a neoliberal economic agenda'.	Leese and Meisch, (2015)
4. It is not a clearly defined construct or an agreed and tested framework [a], not a mature concept and need further improvement [b]. It is promising but still faces significant conceptual and practical challenges [c]	[a] Wichelns (2017), [b] Reinhard *et al.* (2017), [c] Leck *et al.* (2015). Albrecht *et al.* (2018)
5. There is still no common definition, framework and methodology for nexus research	Allouche et al. (2014). *Endo et al.* (2015), Benson *et al.* (2015). Albrecht *et al.* (2018)
6. Not fully acknowledged on the ground and lacks publicity due to under-represented of private sectors and media in its activities	Endo *et al.* (2017)
7. A buzzword derived from an ambiguous meaning and strong normative resonance. The usage of this term are plural, fragmented, and ambiguous in the UK natural resource debates	Cairns and Krzywoszynska (2016)

Main criticism	Reference
8. It is still an evolving concept that has remained largely in the conceptual realm [a] with a high-level insights [b]	[a] (Smajgl et al., 2016) [b] (Galaitsi et al., 2018)
B. The application of WEF nexus approach	
1. Policy processes may lead to delays [a], slowness, and inertia [b] in policy-making process, especially in developing countries	[a] Wichelns (2017), [b] Mitchell et al. (2015)
2. It is not only involves technical issues but also political issues [a] while nexus concept is often inconsistent in politics of sustainability [b]. In addition, it mostly depoliticised, neglecting historical, social and political treatment [c]	[a] Middleton et al. (2015), [b] Leese and Meisch, (2015), [c] Foran (2015)
3. In site-specific studies and the vertical integration of local nexus issues within national and global nexus issues was often missing [a] due to the complex nature of the nexus [b]	[a] Endo et al. (2017), [b] Allouche et al. (2014)
4. Nexus analyses are insufficiently cross-sectoral, focusing mostly on water that assigns unequal sectoral weighting	Smajgl et al. (2016)
5. There is no clear boundaries to constrain the WEF nexus applications.	Galaitsi et al. (2018)
6. Some of existing models and frameworks failed to capture interconnections among variables due to lack of data sharing and availability	Shannak et al. (2018)
C. Outcome and impact of WEF nexus approaches	
1. The research on the nexus influence on the decision making by stakeholders are limited [a] and continues to fall short of expectations of its research-backed benefits [b]	[a] Wichelns (2017), [b] Dargin et al. (2019)
2. The current WEF nexus discourse fails to adequately consider the politics inherent in WEF sector	de Grenade et al. (2016)
3. There is lack of evidence from WEF nexus research that has produced an intellectual toolkit including validated claims showed the improvement of resource management and governance outcomes.	Galaitsi et al. (2018)
4. There is a dearth of WEF nexus adoption in national policies, programmes and institutions. The gender aspects are also often overlooked in WEF nexus assessments.	FAO (2018)
5. Ignorance of the main democratic goal of sustainability concept through over-emphasizing resource security level at the expense of livelihoods	Biggs et al. (2015)
6. Operationalising WEF nexus is suggested in many studies and is urgently needed to bring the 'nexus thinking' into 'nexus doing'	Simpson and Jewitt (2019)

2.4 EFFORTS TO NARROW GAPS AND ADDRESS CRITICISMS

From the analysis in Sections 2.2 and 2.3, common nexus research gaps have been identified in both the frameworks and in nexus critiques and reviews. This section presents recent work that has started to address some of these gaps.

2.4.1 Narrowing knowledge gaps

An overview of studies that start to address gaps is compiled in Table 2.4. This table distinguishes three main knowledge gaps, namely: (1) lack of WEF-related datasets and knowledge; (2) insufficient availability of applications; and (3) lack of agreement and clarity of several WEF-related issues. These are discussed in more detail.

There are several data that are still considered poorly covered or poorly available for WEF analyses. Two examples include aquifer data in water-scarce regions and consumptive water use data in the energy sector (Macknick *et al.*, 2012), which are elaborated in more detail here. The five studies in Table 2.4 (section A1) start to contribute in filling the first example of aquifer data. MacDonald *et al.* (2012) employed GIS-based analysis to establish continent-wide maps of aquifer storage and potential borehole yields in Africa. By reviewing maps, data, and publications, quantitative maps of groundwater in Africa can be developed to assess water security at the national and regional levels. Lezzaik and Milewski (2018) have attempted to deal with the paucity of aquifer data in MENA regions. They estimated groundwater storage reserves based on saturated thickness and effective porosity estimates of groundwater using GIS-based models. Additionally, to measure the alteration in groundwater storage, monthly gravimetric datasets (GRACE) and land surface parameters (GLDAS) were used. Modelling approaches such as MODFLOW (van Camp *et al.*, 2013), random forest models and maximum entropy models (Rahmati *et al.*, 2016), water balance equations and water table fluctuation analysis (Rezaei and Mohammadi, 2017) have been used to estimate aquifer yields in several water-scarce regions such Iran, Libya, Egypt, Sudan, and other African countries.

Studies on the consumptive water use in the energy sector (Table 2.4, A2) are more numerous. For instance, Davies *et al.* (2013) investigate consumptive water demand for electricity production at the global level using an integrated assessment model of energy, agriculture, and climate change (the GCAM model). Similarly, Mekonnen *et al.* (2015) evaluate the global consumptive water footprint (WF) of electricity and heat generation in the fuel supply, construction and operational stages. Other researchers applied similar objectives to different regions, such as in China (Liao *et al.*, 2016) and the European Union (EU) (Vanham *et al.*, 2019). A review paper by Dodder (2014) highlighted some scenarios of future water demand in the energy sector.

Discussions on the impacts of hydropower and other water resources developments on aquatic ecosystems and full life-cycle assessments in terms of water and energy impacts have been addressed by various studies (Table 2.4, A3 and A4). The gaps addressing studies on topics A5 and A6 (Table 2.4) are insufficiently addressed. For example, the topic of nutritional water productivity is still very limited, although a few studies have explored the issue (Nyathi *et al.*, 2016; Nyathi *et al.*, 2018; Nyathi *et al.*, 2019; and Nouri *et al.*, 2020). These studies contribute to narrowing knowledge gaps in the WEF nexus, especially in relation to ecosystem services (cf. Hülsmann et al., 2019). Similarly, the discussion on energy productivity in agriculture (A6) has not been ad-dressed adequately, though some studies have attempted to deal with this issue e.g. Ball *et al.* (2015); Moghaddasi and Pour (2016); Elsoragaby *et al.* (2019); Rautaray *et al.* (2020).

Table 2.4. Summary of research related to addressing WEF nexus knowledge gaps

Hoff's knowledge gaps*	Related studies
A. Lack of WEF datasets & knowledge on:	
1. the available water resources data especially on safe aquifer yields in 'economically water scarce' regions	(MacDonald *et al.*, 2012); (van Camp *et al.*, 2013); (Rahmati *et al.*, 2016); (Rezaei and Mohammadi, 2017); (Lezzaik and Milewski, 2018)
2. the consumptive water use in the energy sector, compared to withdrawal data	(Davies *et al.*, 2013); (Dodder, 2014); (Mekonnen *et al.*, 2015); (Liao *et al.*, 2016); (Pan *et al.*, 2018); (Vanham *et al.*, 2019)
3. the impacts of hydropower and other water resources development on aquatic ecosystems	(Liermann *et al.*, 2012); (Odiyo *et al.*, 2012); (Elosegi and Sabater, 2013); (Yan *et al.*, 2015); (Fan *et al.*, 2015); (Hecht *et al.*, 2019)
4. the full life-cycle assessments in terms of water and energy	(Feng *et al.*, 2014); (Al-Ansari *et al.*, 2015); (Pacetti *et al.*, 2015); (Mannan *et al.*, 2018); (Masella and Galasso, 2020)
5. water productivity per nutritional content of food products	(Nyathi *et al.*, 2016); (Nyathi *et al.*, 2018); (Nyathi *et al.*, 2019); (Nouri *et al.*, 2020)
6. energy productivity in agriculture	(Ball *et al.*, 2015); (Moghaddasi and Pour, 2016); (Elsoragaby *et al.*, 2019); *(Rautaray et al.*, 2020)
B. Insufficient availability of:	
1. the uniformly applicable 'water footprint' framework regarding water use efficiency for different forms of energy or food production	(Okadera *et al.*, 2015); (Wang *et al.*, 2014); (Hoekstra, 2017); (Ababaei and Ramezani Etedali, 2017); (Das *et al.*, 2020); (Zhai *et al.*, 2021)
2. the harmonized 'nexus database' or analytical framework for monitoring or trade-off analyses	(Howells *et al.*, 2013); (McCarl *et al.*, 2017); (Sušnik *et al.*, 2018); (Lawford, 2019); (Purwanto *et al.*, 2020a); (Nhamo *et al.*, 2020); (Sadegh *et al.*, 2020)

Hoff's knowledge gaps*	Related studies
3. the blueprint for overcoming institutional disconnect and power imbalances between sectors	(Villamayor-Tomas *et al.*, 2015); *(Artioli et al.,* 2017); (Weitz *et al.*, 2014); (Märker *et al.*, 2018); (Mercure *et al.*, 2019); (Pahl-Wostl, 2019); (Bréthaut *et al.*, 2019)
C. Lack of agreement and clarity on:	
1. the water quality standards for different crops and production systems	(Hooda *et al.*, 2000); (Love and Nejadhashemi, 2011); (Allende and Monaghan, 2015); (Chalar *et al.*, 2017)
2. the impact of policy frameworks on water and energy use and resource use efficiency in food production	(Ringler *et al.*, 2013); (Karnib, 2018); (van Gevelt, 2020); (Wu *et al.*, 2021)
3. the impacts of increasing energy scarcity on water and food security	(Ahmad and Khan, 2017); (Dinar *et al.*, 2019); (Liu and Chen, 2020)
4. how to deal with the increasing level of complexity that comes with higher levels of integration	(Wichelns, 2017); (Altamirano *et al.*, 2018); (Shannak *et al.*, 2018); (Albrecht *et al.*, 2018);(Dargin *et al.*, 2019); (Mercure *et al.*, 2019)

*Adapted from Hoff (2011)

Not all studies directly address Hoff's knowledge gaps. Regarding the issue of energy productivity (A6), most studies link energy productivity with the industrial or manufacturing sectors, but neglect the agricultural sector, possibly omitting an important source of energy demand and production. As another example, there is considerable research related to the water footprint of energy and food production (see B1), but not all studies use a consistent framework, thereby precluding common assessment and comparison. The availability of a harmonized database and indicators for the WEF nexus, an analytical framework that is able to monitor the potential trade-offs and synergies in WEF resource management, and WEF nexus analysis that can resolve institutional disconnects and power imbalances are the main gaps that must be prioritized to be addressed in the future nexus studies and applications.

2.4.2 Addressing criticisms in WEF nexus frameworks

Further issues are suggested that should be considered as improvements of WEF nexus research to assisting planners and policy-makers. Three main issues are identified: (1) participatory stakeholder involvement in designing and carrying out nexus research; (2) a comprehensive, open–access (where possible) WEF nexus database; and (3) an updated WEF nexus framework to support policy and decision-making, including the concept of WEF resource security, which is rarely considered, although it is a central aspect (Purwanto *et al.*, 2020a; Martinez *et al.*, 2018).

The first issue regards participatory engagement of related stakeholders to raise awareness and increase the understanding of the nexus for those responsible for its planning and management e.g. Wichelns (2017), Sperling and Berke (2017), Albrecht *et al.* (2018), Shannak *et al.* (2018), and Simpson and Jewitt (2019). Sušnik *et al.*, (2018) engaged multiple stakeholders from project inception to development a WEF nexus serious game to understand the interaction between water, energy, food, land, and climate. Through the SIM4NEXUS project (Ramos *et al.*, 2020), which includes 12 case studies from regional to global scale, stakeholders and local partners were involved in all nexus modelling stages including conceptualization, quantita-tive model development, validation and implementation of serious games. Using another approach, Purwanto *et al.*, (2019) implement a group model building (GMB) technique to develop qualitative causal loop diagrams of WEF nexus security in Indonesia by involving local expert stakeholders. The awareness raising and better understanding of stakeholders about the complexity of the WEF nexus was one of the main outcomes in that process. A study on early stakeholder involvement to ensure perspective convergence among researchers and stakeholders in WEF-related sectors was conducted by Daher *et al.* (2020) in the San Antonio Region. They provided questionnaires to 370 respondents from three different groups (i.e. government institutions, non-government/non-profit, and business enterprises). The main objectives were to evaluate the level of convergence of nexus understanding and to identify barriers and opportunities to improving communication among stakeholders. There are many studies that have applied participatory approaches in WEF research, including fuzzy-cognitive mapping, online investigation and snowball sampling (Martinez et al., 2018), interviews, focus group discussions and vision-building workshops (Mguni *et al.*, 2020), and multi-objective optimization methods for WEF nexus management and the involvement of multiple stakeholders (Cansino-loeza and Ponce-ortega, 2020). These studies demonstrate that multiple actors should work together for continuous improvements to make the nexus approach useful in the planning, evaluation and decision-making processes. Researchers are involving key stakeholders throughout the nexus investigation process more regularly.

To make the WEF nexus concept work for quantitative assessment, valid, integrated, and open data sources at all levels and scales must be available for governments and scientific institutions. As listed in Table 2.4 (B2), this is crucial because current WEF nexus data availability can still pose a major challenge in the analytical process, especially at the local level (Hoff, 2011; Liu *et al.*, 2017; Shannak *et al.*, 2018). The problem of data availability, validity, quality, and accessibility in many countries particularly in developing countries is common. Several global, regional and country-level WEF data sources are more readily available compared with data at the local level. However, existing datasets are not comprehensive, data quality and reliability can be questionable, and data are separated from each other and not contained in a 'WEF system database'.

Table 2.5 summarizes some of the sources of WEF-related data at global, regional, and country levels.

Table 2.5. Some water, energy, and food data sources

Data source	Type of data	Level
Water		
• FAO (Food and Agriculture Organization) – AQUASTAT (http://www.fao.org/nr/water/aquastat/data/)	• Land use, economy-development-food security, precipitation, renewable water resource, dam capacity, water withdrawal, wastewater, irrigation and drainage, water conservation, water harvesting, flood occurrence, drinking water access	• Global • Country
• Water Footprint Network (*https://waterfootprint.org/en/resources/waterstat/*)	• Product water footprint, national water footprint, International virtual water flow, monthly gridded blue water footprint, water scarcity, water pollution level	• Global • Country
• USGS data (*https://waterdata.usgs.gov/nwis*)	• Surface water, groundwater, water quality, water use	• Global • Country
Energy		
• IEA (International Energy Agency) • (*https://www.iea.org/data-and-statistics*)	• Energy supply, energy consumption, electricity, energy import-export, CO_2 emission, energy prices, renewable energy	• Global • Regional • Country
• IRENA (International Renewable Energy Agency) (*https://www.irena.org/Statistics*)	• Capacity and generation, energy balances, energy transition, energy policy, cost, climate change, finance and investment	• Global • Regional • Country
Food		
• FAO (Food and Agriculture Organization) (*http://www.fao.org/faostat/en/#data*)	• Food production, trade, food balance, food security, price, inputs, population, investment, macro-statistics, agri-environmental Indicators, emission-agriculture, emission-land use, forestry	• Global • Country
• OECD-FAO (*http://www.agri-outlook.org/data/*)	• Agricultural product, consumption, imports, dairy, meats, fishery, etc.	• Global • Regional • Country
Multiple data		
• (*http://data.worldbank.org/indicator/*)	• Data and indicators related to agricultural data, economy, energy, environment, climate change, water etc.	• Global • Country
• (*https://ourworldindata.org/energy*)	• Agricultural production, meat & dairy, fishery, energy, access to energy, renewable energy, air pollution, clean water, sanitation, etc.	• Global • Country

One significant effort is the development of an integrated data and analysis toolbox called NeFEW (Nexus of Food, Energy and Water) to incorporate available global datasets (Sadegh *et al.*, 2020). This toolbox gathers data to allow for modelling and analysis of WEF resources and their interconnectedness at a country level. Lawford (2019) analysed the importance of integrated WEF nexus data and information to assist practitioners in planning and decision-making processes. He proposed WEFDIS (WEF nexus data and information system) to ensure that WEF data and information from satellites, in-situ data networks, and other data sources are readily available. Eight proposed sequential and parallel measures to develop and implement WEFDIS and to structure WEFDIS are discussed in Lawford (2019). Accessibility and standardization of data are two important points to communicate and consolidate data and information from various sources. The incorporation of existing WEF-related databases is a good example of potential data integration and would enable replication at a smaller scale. To harmonize databases at a local level, more research appears to be needed, largely due to the diversity of data sources and formats.

On-the-ground WEF nexus implementation is the third issue. Operationalizing the WEF nexus to assist policy-makers and other stakeholders in managing resources is a main recommendation in several WEF nexus-related discussions, including Shannak *et al.*, (2018), Albrecht *et al.*, (2018), and Simpson and Jewitt, (2019), and is included in Hoff's knowledge gaps (Table 2.4, B3). The improvement of the nexus concept by transitioning towards 'doing' instead of only 'thinking' has been established in several studies. Studies by Purwanto *et al.* (2019; 2020a) are examples on how to integrate the WEF nexus concept into local planning systems to assist local stakeholders in achieving WEF security targets in a regional context. Other evidence of WEF nexus implementation can be seen in Hoff *et al.* (2019) through five case studies in MENA countries (Jordan, Lebanon and Morocco). They evaluate the current nexus conditions and to link that to Sustainable Development Goal (SDG) ambitions and Nationally Determined Contributions (NDCs). An integrative study to assess the progress towards SDGs targets in South Africa using a WEF nexus analytical model has been conducted by Nhamo *et al.* (2020) who applied seven WEF nexus composite indices (water availability, water productivity, energy accessibility, energy productivity, food self-sufficiency, cereal productivity, and an integrated WEF index) to evaluate SDG targets 2, 6 and 7 in the period 2015 to 2018. Further research is expected to strengthen the concept and bring the new methodologies and empirical evidence to influence policy and decision-making processes *(Brouwer et al.*, 2018).

Regarding responses to criticisms of the WEF nexus approach in general, one valuable commentary comes from Brouwer et al. (2018), in response to Galaitsi *et al.* (2018). They offer evidence from the Horizon 2020 SIM4NEXUS project (www.sim4nexus.eu) that identifies the added value of the WEF nexus concept, including flexibility and adaptability, the ability to identify critical nexus-relevant policy objectives, and the better identification of trade-offs and synergies for resource management and policy-making. They contend that the nexus concept is supportive in WEF-related policy-making processes. More specifically, three key features of the nexus concept were proposed to be improved upon, namely: (1) focus on bio-physical, socio-economic and policy interactions; (2) seeking a balance between different needs to achieve sustainable and integrated natural re-sources management; and (3) a systematic effort for policy coherence across sectors.

2.5 UPDATING THE WEF NEXUS FRAMEWORKS

Despite the progress made in addressing nexus research gaps, some areas still remain unaddressed. In this section, four main issues still needing research are proposed, and these are incorporated into an updated WEF nexus framework. These issues are (Figure 2.3):

(1) Making the nexus relevant for stakeholders and policy

This underlines the importance of participatory engagement to ensure that stakeholders in the water, energy, and food sectors can understand the interlinkages in the nexus and what this means for policy and decision making. Several methods could be used such as participatory modelling, group model building (cf. Purwanto et al. 2019), focus group discussions, and surveys and interviews, but inclusion of relevant stakeholders throughout is critical.

(2) The issue of reliable data and information

Any WEF nexus study outputs should be based as much as possible on reliable data that are valid and integrated, and that are available with a good level of accessibility to facilitate quantitative analysis and providing robust, defensible results. Ideally, a universal, open-access, WEF database would be developed.

(3) Creating an adaptable framework

Framework adaptability is important in the WEF nexus due to the diversity of resources, natural conditions, scales, levels, government and planning systems, the responsibility of institutions, laws and regulations, and key nexus foci. As such, any framework must have the flexibility to adapt to a diverse set of circumstances.

(4) Be easily applicable

Incorporating the WEF nexus into planning and decision-making systems is essential to move the WEF nexus from a concept to an operational framework that brings real benefits for a more sustainable and integrated policy-making process.

To include these principles, a bottom-up "local to global" approach is proposed so that lower levels of government are able to identify WEF resource challenges and solve them through proper planning and actions, using locally relevant approaches exploiting the best available data and information. Well-achieved targets at the local level may add-up to affect WEF resource security at a higher level of governance. However, the four challenges can also refer to what is done at a higher level. For example, the availability and integration of WEF datasets at the global and national levels is relatively better than the local level. Therefore, methods to develop indices, parameters, and individual datasets can be adopted from global and national sources for local scales, thereby potentially improving multi-scale nexus assessment and relevance.

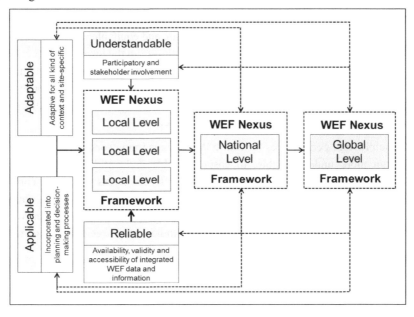

Figure 2.3. The main proposed principles and perspective for future WEF nexus concepts and frameworks

2.6 CONCLUSION

This chapter has investigated knowledge gaps, criticisms, and areas for improvement related to research on the WEF nexus that have emerged since the concept of WEF (security) nexus was proposed. Thirteen knowledge gaps were identified by Hoff (2011). During the last decade, a number of efforts have been made to narrow these gaps. In this study, 67 papers are reviewed that consider Hoff's gaps. Despite significant progress, some gaps have not been entirely fulfilled, such as those related to energy productivity in agriculture, harmonizing a WEF nexus database, and the relevance of WEF resource security. Critiques on the concept, application, and operationalisation of the WEF nexus approach have been considered in this study.

Furthermore, ten WEF nexus frameworks have been analysed and compared with sixteen suggested features for WEF nexus inclusion. The following issues are insufficiently addressed in existing frameworks: the WEF security focus (resource availability, accessibility, and quality); robust systems approach and datasets; participatory stakeholder engagement in nexus research. Additionally, a general lack of the importance of ecosystems and their services is prevalent. Local perspectives are often under-represented, especially in developing countries with decentralized governance systems. Context specific practical and policy implementation guidance in evaluation and planning still needs to be improved. It is suggested that locally-based WEF management will help ensure that WEF re-sources are managed in a holistic and equitable way. The WEF nexus approach should move from thinking to doing, starting from the lowest level (i.e. moving from conceptual ideas to practical and relevant applications). Stakeholder participation is crucial to man-age WEF resources. At the same time, it is critical to prevent delays in the process of decision-making that can be caused by ineffectiveness of time allocation to accommodate the various kinds of stakeholder's interests.

Four principles and perspectives for future WEF nexus framework development (i.e., to make them more understandable, to ensure reliable and valid data, to make them adaptable to many diverse situations, and to be applicable across scales) are considered central to increasing the benefits and improving the role of the WEF nexus concept in influencing policy and resource planning processes. Continuous improvements, especially in grounding the WEF nexus concept, indicate the urgent challenge to better manage the three resources of water, energy and food.

3

WEF-RELATED SECTORS IN LOCAL ECONOMIC DEVELOPMENT

Abstract

Water, energy, and food (WEF) related sectors are important to support people's life in a region. Resource evaluation is one of the stages in resource management to ensure that the existence of those sectors is provided sustainably. The assessment of the agglomeration level and growth of each sector in economic development can give better insights for local stakeholders either government bodies or private firms to improve sustainable management of these sectors. The objectives of this chapter are to portray the agglomeration level and recent growth of WEF related sectors in local regions in Indonesia and to determine possible sustainable development strategies. The location quotient (LQ) and competitive position (CP) analysis methods are employed in this regard. By analysing Gross Regional Domestic Product (GRDP) between 2000 and 2015, basic and non-basic sectors have been determined. Results show that the general characteristics of WEF related sectors in this region can be distinguished clearly based on its main economic development focus. Results show recent growth in WEF sectors locally, from which possible strategies for future sustainable development are formulated that could be considered in the evaluation and planning process. This approach can be expected to assist local government and stakeholders in undertaking preliminary evaluation, in particular the availability of WEF resources, ensuring that development meets local and national sustainable development targets.

Keywords: location quotient, competitive position, basic sector, non-basic sector, water-energy-food (WEF), gross regional domestic product (GRDP)

This chapter is based on:

Purwanto A., Sušnik J., Suryadi F.X., de Fraiture C., (2018), *Determining strategies for water, energy, and food-related sectors in local economic development*, Sustainable Production & Consumption 16, 162–175, Elsevier, DOI: https ://doi.org/10.1016/ j.spc.2018.08.005 (Published).

3.1 INTRODUCTION

Equitable access to, and proper management of water, energy, and food (WEF) sectors play an important role in determining the efficacy of poverty alleviation, welfare improvement of people in a region, and sustainable development (FAO, 2014) and is essential for human life. Availability, accessibility, and quality of water, energy, and food are the primary components of these resources that should be addressed in an integrated manner. Inward-looking and resistance to sharing information and resources among departments concerning water, energy, and food sectors leads to ineffective achievement of national and local sustainability targets. Based on global projections for the next decades, Hoff (2011) concludes there will be a prominent increase of water, energy, and food demand due to population growth, economic activities, changing in diets, culture, technology, and climate. Global population increased by two billion during the period 1990–2015. At present, one in nine people has insufficient food, while one third is malnourished (UNDP, 2016). There are around 750 million people with lack of improved drinking water access, while global industrial water demand is predicted to escalate by 400% during 2000–2050 (UN-WWAP, 2015). Additionally, energy demand will almost double, while the demand for water and food are foreseen to escalate by more than 50% by 2050 (IRENA, 2015).

The concept of sustainable development is relevant with respect to water, energy, and food issues. Not only should these sectors be secured (availability, accessibility, quality), but this should be achieved in a sustainable way. In 1987, the Bruntland Commission Report provided a definition of sustainable development as 'development that meets the needs of the present without compromising the ability of future generations to meet their own needs' (United Nation, 1987). In this context, (Endo et al., 2015) explain that interlinkages between water, energy and food are very complicated and have become crucial for the global community in the future to handle this issue in a sustainable way. Additionally, Hoff (2011) underlines that resources availability is not the one and only factor of security. Accessibility and quality are crucial to be covered in achieving resource security. Focusing only to certain aspect and resource, without considering others will potentially cause conflict and unbalanced competition among sectors (i.e. the current 'nexus' approach). The long-term stability of resources is one of the final goals of sustainable development. It only can be achieved by integrating and acknowledging social, environmental, and economic aspects in decision-making process (Emas, 2015). Those three aspects have to be evaluated and planned correctly to ensure the achievement of the development goals in each region. In this context, the evaluation of a WEF-sector's agglomeration or concentration and its trends in a region are significantly important to give better understanding and insights for local government and stakeholders in setting the goals and allocating budget for development activities. By assessing recent trends and potential future directions, the sustainable development of these resources can be better guided, especially in the context of national and local sustainability targets. Sector agglomeration defines how concentrated a sector is within a region. It is important for those sectors which greatly affect people's lives such WEF-related sectors. One of the perspectives that can be used to do this kind evaluation from an economic point of view is economic base analysis (EBA). EBA is crucial for governments to understand better those sectors that contribute disproportionally to the economic growth of a region. This theory has prolonged practice regarding planning and geography with the main assumption that all economic activities in a region can be classified into basic and non-basic industries or activities (Isserman, 1977; Wang and Hofe, 2007). Furthermore, Wang and Hofe (2007) proposed one of the assumptions related to basic and non-basic sectors, where basic sectors yield more goods or services than the local needs, so the surplus can be exported to other regions. On the contrary, the non-basic sector is assumed under the level of self-sufficiency and therefore unmet demand needs to be imported.

One common method in economic base analysis is Location Quotient (LQ). This technique has been employed in many sectors such as the mapping crime (Brantingham & Brantingham, 1998), trade sector (Chiang, 2009), industrial concentration (Billings & Johnson, 2012), and carbon emission (Trappey *et al.*, 2013), marine sector (Morrissey, 2014), economic development (Alhowaish, 2015), and road project development (Berawi *et al.*, 2017) among others. In addition, there are many studies employing the competitive position method to analyse various sectors (for instance Horta and Camanho, 2014; Dang and Yeo, 2017). However, although extensive research has been carried out on sector's agglomeration analysis, no single study exists which focuses on water, energy, and food-related sectors in more detail and in the perspective of local economic and sustainable development. In this respect, this study is novel, and could yield new insight for policy makers, especially when developing sustainability targets, and when evaluating progress towards existing targets.

This study analyses the agglomeration of WEF related sectors and other sectors in three characteristic local regions in Indonesia and WEF related sub-sectors in an agriculture–manufacture based region. This research also determines strategies for WEF related sectors based on agglomeration and competitiveness. This work analyses gross regional domestic product (GRDP) in the year 2011–2015 and 2000–2015 using the combination of LQ techniques and competitive position charts (Zhao *et al.*, 2016).

3.2 METHODS

3.2.1 Dataset

The data used in this study were taken from Statistics of West Java report, Statistics of Karawang Regency, Statistics of Cianjur Regency, Statistics of Bekasi City, and other related data sources. The main datasets used in LQ computation are all at constant market price (CMP) as follows: (1) GRDP of West Java Province at 2000 CMP by industrial origin (2000–2015); GRDP of Karawang Regency at 2000 CMP by industrial origin (2000–2015); GRDP of West Java Province at 2010 CMP by industrial origin (2011–2015); GRDP of Karawang Regency at 2010 CMP by industrial origin (2011–2015); GRDP of Cianjur Regency at 2010 CMP by industrial origin (2011–2015); GRDP of Bekasi City at 2010 CMP by Industrial origin (2011–2015).

The data used in this study were the statistical data series of 16 years of GRDP (2000–2015) that has been verified and analysed by the statistical agencies. Gross Domestic Product (GDP) or Gross Domestic Regional Product (GRDP) at the provincial/regency level are arranged based on a production and expenditure approach. Production-based GRDP by industrial origin denotes a basic measure of value-added emerging from various kind of economic activities and production in a region. GRDP therefore implicitly accounts for the basic factors of production in its metric, but does not report on the effects of these factors explicitly. It has changed since 2010 in the number of categories from 9 main sectors to 17 main sectors, comprises 39 sub-sectors, and 5 sub sub-sectors.

In this study, the water-related sector is based on the definition from statistics agencies which consists of domestic and industrial water supply, sewerage, waste management, & remediation products and services. Agricultural water or irrigation purposes for instance is not specifically included in this statistical value. In addition, the term 'water supply' was used in the period of 2000–2010 and classified as 'sub-sector'. Meanwhile, in the period of 2011–2015, the 'water supply, sewerage, and waste' term categorized as 'sector'. Both terms have similar definition and coverage. Therefore, to avoid confusion, the term 'water supply' is used in this study to represent the water sub-sector. The energy-related sector covers electricity, manufacture of gas, and production of ice. Furthermore, the food-related sector is defined as agriculture, livestock, hunting, & agriculture services (with sub-sectors; food crops, horticultural crops, plantation crops, livestock, agriculture services & hunting), forestry and logging, and fishing. All the definitions and coverage may slightly differ between authorities or government institutions, however there basic levels of coverage remain similar and are considered comparable for this study.

3.2.2 Location quotient (LQ)

There are four basic techniques in economic base analysis that can be used, especially if the availability of data becomes an obstacle in using more complex economic analysis. These are: survey methods, assumption methods, location quotient (LQ) method, and minimum requirement method (Wang & Hofe, 2007). According to (Leigh, 1970), the best measure to determine basic and non-basic sector is the use of primary surveys. Nevertheless, this can be very difficult and costly, in particular for a large region with diverse economic activities (Leigh, 1970; Brodsky & Sarfaty, 1977). LQ is the methodology adopted for this study. Miller *et al.* (1991) defined LQ as 'a basic analytical tool to yield a coefficient or simple expression of how well represented a particular industry is in a given study region'. LQ has been widely applied in economic geography and regional economics since the 1940s due to the data unavailability (at that time) of interregional trade flow (Wang and Hofe, 2007). It can be applied to compare the role of industrial sectors in a region with the same variable in the higher regional level to understand local potential on basic and non-basic sectors. LQ is still applied in the regions with relatively little economic interregional data are available (Chiang, 2009; Day and Ellis, 2012; Islam *et al.*, 2016).

Several questions and criticisms have been raised about the accuracy of LQ method in estimating regional economic impact (Leigh, 1970; Tohmo, 2004; Riddington *et al.*, 2006). Nevertheless, this technique has remained a highly popular due to its simplicity (Miller *et al.*, 1991), non-intensive data requirements, analytical skill, time efficiency to compute, low budget requirements (Isserman, 1977), and the lack of interregional trade flow and primary data requirements (Richardson, 1985). Suyatno (2000) used the dynamic location quotient (DLQ) in combination with static location quotient (SLQ) to further analyse the changes or sectoral reposition of each sector by considering GRDP and the annual growth rate of each sector at selected years. The basic formula of LQ to calculate the agglomeration level in this study is as follows:

$$LQ = \frac{X_{in}/Y_n}{X_i/Y} \qquad (3.1)$$

where LQ is the location quotient value, Xin represents GRDP Sector *i* in regency/city n, and Y_n is the total GRDP in regency/city n. Afterwards, X_i is GRDP Sector *i* in provincial level, while Y indicates the total GRDP in provincial level. If the value of *LQ* for a given sector is greater than or equal to 1 (*LQ*≥1), it can be classified as Basic Sector, while if the value of LQ is less than 1 (*LQ*<1), it can be categorized as Non-Basic Sector (3.1). Furthermore, the value of agglomeration growth rate (P) is obtained by subtracting the value of agglomeration level of sector *i* at selected year n (LQ_{itn}) from the value of agglomeration level of sector *i* at initial year (LQ_{it0}), divided by the value of LQ_{it0}, then multiplied by 100 percent (eq. (3.2). It indicates the agglomeration growth during the given period. If the value of P is positive and greater than 10% or 0.1 (P>0.1), it reflects that sector *i* is growing and the advantage level of the cluster in the region is increasing. Contrarily, if the value of P is negative and less than −10% or −0.1 (P<−0.1), it means that the growth is declining and the cluster advantage is decreasing. Additionally, if the shift is between +/−10% or +/−0.1 can be considered as very small change.

$$P = \left(\frac{LQ_{itn} - LQ_{it0}}{LQ_{it0}}\right) x100\% \qquad (3.2)$$

3.2.3 Competitive position (CP)

A clear way to depict the value of LQ and its growth effectively is in a bubble chart with four quadrants, the so-called Porter's Cluster (Goetz *et al.*, 2007) or Competitive Position Matrix (Zhao *et al.*, 2016). The competitive position is defined as the relative position of a sector against other sectors within the selected composite indicators (Simmonds, 1986). The horizontal axis in this study reflects the value of agglomeration level (LQ) for the year of analysis, while the growth of LQ overtime (P) is captured by a vertical axis. Bubble sizes represent the relative size of the industry, and clearer pictures of potential economic clusters can be obtained using this composite method (see Figure 3.1).

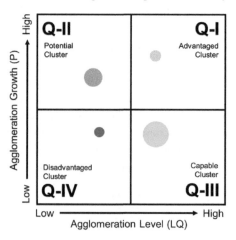

Figure 3.1. Archetype competitive position chart (source: Modified from Zhao et al., 2016)

Quadrant I (Q-I) represents industries that have high agglomeration level and high agglomeration growth so-called advantaged or stars group. Industries in this cluster are strong, advancing, and can be expected to become more dominant in the future. Potential or emerging industrial groups which have low concentration but high growth are indicated in quadrant II (Q-II). This position contains clusters that will be shifting ultimately to the first quadrant under continuous growth in the future. The third quadrant (Q-III) reflects clusters concentrated in the region but that may be declining in importance. These are the so-called capable or mature group. The movement to the less concentrated clusters are likely in the forthcoming years. Clusters with a low concentration and lack of competitiveness are denoted in quadrant IV (Q-IV) and named disadvantaged or transforming group. All the GRDP data were analysed using LQ methods, then visualized in the competitive position bubble chart. Finally, strategies for water, energy, and food-related sectors and sub-sectors can be determined based on their agglomeration, competitiveness, and growth.

3.3 STUDY AREA

Karawang Regency and two others local regions have been chosen to represent three characteristics of economic development in Indonesia. All regions are located in West Java Province. Karawang Regency is the main focus of analysis with agricultural and manufacturing activities as its major economic development activities. Cianjur Regency is an agricultural-based region, while Bekasi City specialises in trade, services, and manufacture (Figure 3.2). West Java province has a population of 46.7 million inhabitants (2015 statistics). The total area is 35,377.76 km^2, and recently it is divided into 18 regencies and nine cities. About two-thirds (66.5%) of people in this province live in urban areas. In terms of economic sectors, 18.79 million workers are employed in trading, manufacturing and other sectors with the composition of 27.1%, 21%, and 19.2% respectively. By 2015, the economic structure of this province was dominated by processing or manufacturing industry (43%), trading (15%), agriculture (9%), construction (8%), and other sectors (25%) (BPS of West Java, 2016a).

Karawang Regency is also one of the largest agricultural centres in Indonesia with paddy as the main crop. On the other hand, this region focuses on industrial development as stated in its long-term (twenty-yearly) planning year 2005-2025. Its vision is to achieve a prosperous region based on agricultural and industrial development. However, in terms of sustainable development, these targets seems at odds with each other. Karawang comprises irrigated paddy field 97,353 Ha (50.7%), non-irrigated agriculture area 10.5 Ha (5.5%), fish pond and aquacultures 18.79 Ha (9.8%), and non-agricultural area such as buildings, settlement, industry, road and water bodies 44.9 Ha (23.4%) (BPS of Karawang, 2015). The main commodities of horticulture products are cucumber, beans and mushroom, whereas for livestock, cows, buffalos, sheep, goats and chickens are raised in this area. There are also another food sources such as capture fisheries (8,591 tons), aquaculture (42,483 tons), and salt production (8,446 tons)(BPS of Karawang, 2015). By 2016, around 954 units of large manufacturing companies, and approximately 9,290 units of intermediate and small firms already existed in Karawang. Furthermore, the total area provided by the government of Karawang to develop industrial estates is become one of the largest among other regions in South East Asian countries. It indicates that industrialization policy is a major policy direction locally and nationally.

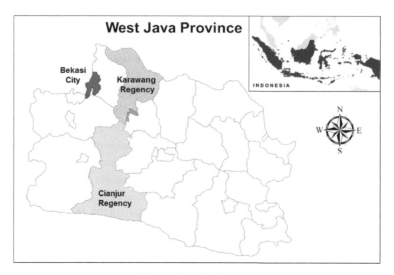

Figure 3.2. Study area

Each study site has its own specialization and strength in particular sectors. There are national targets in water, energy, and food-related sectors stated in national long-term and mid-term planning. However, several acute problems have obstructed the attainment of the targets, i.e. resources mismanagement, lack of coordination, and authority imbalance among sectors, levels and scales (Bellfield *et al.*, 2016). Additionally, resource management that focuses only on a certain sector (silo thinking) is postulated as one of the major causes of ineffectiveness in target achievement both in local and national level. More detailed data and information of each local region that will be discussed in this study area summarized in Table 3.1, while the average of GRDP is presented in Table 3.2.

Table 3.1. Description of the three study areas

Aspect	Karawang	Cianjur	Bekasi
Location	107°02'-107°40'E and 5°56'-6°34'S	106°42'-107°25'E and 6°21'-7°25'S	106°55'E and 6°7'-6°15'S
Elevation	0-1,291 m asl	7-2,962 m asl	11-81 m asl
Slope	0-40%	0-40%	0-2%
Rainfall	(1,100-3,200 mm)	(2,610 mm)	(3,483 mm)
Avg temperature	(27–33°C)	24°C	(24–33°C)
Total area	1,753.27 km^2	3,614.34 km^2	210.49 km^2
Total population*	2,273,579	2,243,904	2,733.240
Population density*	1,297	621	12,985
Population growth*	1.04%	0.38%	2.74%
Paddy field*	96,482 ha	65,782 ha	475 ha
Non-paddy area*	39,402 ha	200,027 ha	
Non-agri area*	39,997 ha	84,309 ha	20,551 ha

Aspect	Karawang	Cianjur	Bekasi
Area division*	30 sub-districts, 309 villages	32 sub-districts, 360 villages	12 sub-districts, 56 villages
Civil servant*	13,571	15,030	12,943
Labour force*	Total labour force 873,847: agriculture (16%), manufacturing (28%), trade (27%), services (15%), others (14%)	Total labour force 863,592: Agriculture (36%), manufacturing (9%), trade (28%), services (11%), others (16%)	Total labour force 1,091,936: agriculture (1%), manufacturing (27%), trade (26%), services (21%), others (26%)
Agricultural commodity	paddy, horticulture	paddy, horticulture, tea, coconut, palm oil, forest product	paddy, horticulture (production is getting lower)
Livestock commodity	cow, sheep, poultry	cow, buffalo, egg, milk	cow, sheep, poultry
Fishery commodity	freshwater, seawater, processed fish product	freshwater, seawater, processed fish product	freshwater fish
Number of manufacturing industry*	large (954 units), intermediate and small (9,290)	large (13 units), intermediate (79), small (46)	large and intermediate (41,694 units)

*Source: *Data 2015 (BPS-Cianjur, 2016; BPS-Karawang, 2016; BPS-Bekasi, 2016; BPS-West Java 2016a)*

Table 3.2. Average GRDP Year 2011-2015 by Industrial Origin at Constant Market Price 2010 in West Java Province, and three study regions

No	Industrial Origin	Average GRDP 2011-2015 (in Billion Rp*)			
		West Java	Karawang	Cianjur	Bekasi
1	Agriculture, forestry, and fishery	91,030	4,607	7,731	325
2	Mining and quarrying	27,585	3,664	70	-
3	Manufacturing	475,203	84,792	1,330	18,364
4	Electricity and gas	5,767	992	20	1,020
5	Water supply, sewerage & waste	845	72	7	41
6	Construction	86,296	4,176	1,930	4,688
7	Wholesale and retail trade	174,354	12,258	4,215	11,731
8	Transportation and storage	48,739	2,006	1,872	4,543
9	Accommodation & foods services	26,262	1,080	1,316	1,733
10	Information and communication	32,402	1,079	730	1,018
11	Financial intermediary services	25,674	1,235	508	1,333
12	Real estate activities	12,486	273	482	845
13	Business activities	4,279	38	150	208
14	Administration & defence, social	23,808	1,122	654	1,031
15	Education	26,353	808	933	955
16	Human health & social work	7,095	252	160	519
17	Other service activities	20,584	914	840	1,366
	Total	**1,088,762**	**119,367**	**22,952**	**49,718**

*Note: *Rp = Rupiah (Indonesian Currency, where US$ 1 = Rp. 13,300, converted in Sep 2017)*
Source: (BPS-Cianjur, 2016; BPS-Karawang 2016; BPS-Bekasi, 2016; BPS-West Java 2016b)

3.4 RESULT AND DISCUSSION

3.4.1 Comparison of sector's agglomeration

In parts 3.4.1 and 3.4.2, results are presented at the sector level of industrial origin such as agriculture, forestry and fishery, electricity and gas, and water supply, sewerage and waste without taking into account the sub-sectors in more detail.

a.   Karawang Regency

Table 3.3 shows the results of LQ and P values of Karawang Regency. During the period 2011-2015, only three sectors in this region were categorized as being basic sector: manufacturing, electricity and gas (energy), and mining & quarrying. With only 3.9% contribution to GRDP and an LQ value of 0.46, agriculture is not classified as a basic sector in this region, and neither is the water-related sector with the LQ value 0.78. The manufacturing sector dominates and has the highest LQ score (1.63), in line with its large contribution to GRDP in Karawang (71.0%). Trading is the second largest sector in terms of contribution to GRDP with 10.3%. A fairly small shift in most sectors in this regency (both positive and negative) is shown with P values from 0.01 to -0.05. The educational sector is growing and the cluster advantage is increasing. In contrast, mining and quarrying and human health are declining, as is the advantage of the clusters. In the long-term planning (2005-2025) of Karawang Regency, its vision is still to incorporate agricultural and industrial development to improve the region's prosperity. From the analysis, it seems that the vision is not compatible with the current situation. The manufacturing sector dominates the GRDP of this region compared to agricultural sector. As one of the satellite cities of Jakarta, the capital of Indonesia, Karawang is shifting from an agricultural to an industrial-based region.

Table 3.3. LQ & P Values of Karawang Regency Year 2011-2015

No.	Industrial Origin	AVG GRDP*	AVG LQ	Category	P
1	Agriculture, forestry, and fishery	4,607	0.46	Non Basic	**-0.05**
2	Mining and quarrying	3,664	1.21	Basic	**-0.11**
3	Manufacturing	84,792	1.63	Basic	0.03
4	Electricity and gas	992	1.57	Basic	0.02
5	Water supply, sewerage & waste	72	0.78	Non Basic	0.02
6	Construction	4,176	0.44	Non Basic	0.08
7	Wholesale and retail trade	12,258	0.64	Non Basic	**-0.02**
8	Transportation and storage	2,006	0.38	Non Basic	**-0.03**
9	Accommodation & foods services	1,080	0.37	Non Basic	0.06
10	Information and communication	1,079	0.30	Non Basic	**-0.05**
11	Financial intermediary services	1,235	0.44	Non Basic	0.01
12	Real estate activities	273	0.20	Non Basic	0.01
13	Business activities	38	0.08	Non Basic	**-0.04**
14	Administration & defence, social	1,122	0.43	Non Basic	0.03

No.	Industrial Origin	AVG GRDP*	AVG LQ	Category	P
15	Education	808	0.28	Non Basic	0.15
16	Human health & social work	252	0.32	Non Basic	**-0.14**
17	Other service activities	914	0.41	Non Basic	**-0.01**
	Total	119,367			

*Source: own analysis (*billion Rupiahs)*

Figure 3.3 indicates the competitive position of each industrial sector in Karawang Regency during the period of 2011 to 2015. The advantaged cluster (quadrant I) consists of two sectors (manufacturing and electricity and gas). The mining and quarrying sector is located in quadrant III, while the remaining sectors are positioned in quadrant II (6 industrial sectors), and quadrant IV (7 industrial sectors). It is interesting to highlight that in a region which is combining two big development issues i.e. agriculture and manufacture development, the agglomeration level and growth of agricultural sector was very far behind the manufacturing sector.

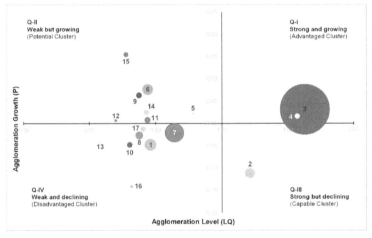

Notes: Numbers in/outside the bubble represent sectors or industrial origin (Table 3.3), while the bubble size indicates the average GRDP year 2011-2015 in Billion Rp

Figure 3.3. Competitive position of each industrial sector in Karawang Regency (2011-2015). Numbers in the bubbles refer to the sectors in Table 3.3.

b. Cianjur Regency

Based on the analysis of LQ and P in Cianjur Regency (Table 3.4), 12 sectors can be classified as basic sector, with agriculture, forestry and fishery sector as the highest score in LQ (4.03). This result highlights Cianjur Regency as an agriculture-based region is still relying on agriculture to develop and export to other regions. Agriculture as the main sector contributes 33.7% of total GRDP of Cianjur. The second largest contributor to GRDP is trading with 18.4%. All sectors are identified to have small change either increasing or decreasing in agglomeration growth (P values ranging from 0.06 to -0.06), except mining and quarrying with the growth of 0.28. Water and energy-related sectors in this region are considered as non-basic sectors with the LQ values 0.41 and 0.17 respectively.

Table 3.4. LQ & P Values of Cianjur Regency Year 2011-2015

No.	Industrial Origin	AVG GRDP*	AVG LQ	Category	P
1	Agriculture, forestry, and fishery	7,731	4.03	Basic	0.06
2	Mining and quarrying	70	0.12	Non Basic	0.28
3	Manufacturing	1,330	0.13	Non Basic	**-0.02**
4	Electricity and gas	20	0.17	Non Basic	0.01
5	Water supply, sewerage & waste	7	0.41	Non Basic	0.01
6	Construction	1,930	1.06	Basic	**-0.02**
7	Wholesale and retail trade	4,215	1.15	Basic	0.03
8	Transportation and storage	1,872	1.82	Basic	**-0.02**
9	Accommodation & foods services	1,316	2.38	Basic	0.01
10	Information and communication	730	1.07	Basic	0.01
11	Financial intermediary services	508	0.94	Non Basic	**-0.06**
12	Real estate activities	482	1.83	Basic	0.01
13	Business activities	150	1.66	Basic	0.02
14	Administration & defence, social	654	1.30	Basic	0.05
15	Education	933	1.68	Basic	**-0.04**
16	Human health & social work	160	1.07	Basic	0.01
17	Other service activities	840	1.94	Basic	**-0.03**
	Total	**22,952**			

*Source: own analysis (*billion Rupiahs)*

The competitive position of each sector in local economic development in Cianjur Regency is shown in Figure 3.4. Eight sectors are located in quadrant I (advantaged cluster), while the rest are located in quadrant II (3 sectors), quadrant III (4 sectors), and quadrant IV (2 sectors). The strong agglomeration of agriculture sectors is clearly depicted. Similarly, the chart reflects the position of the manufacturing sector in Cianjur which was less concentrated and showing a declining trend in agglomeration growth.

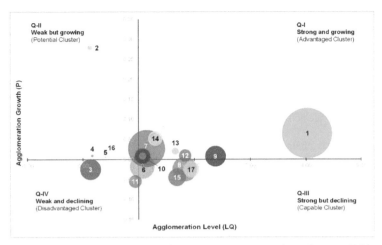

Notes: *Numbers in/outside the bubble represent sectors or industrial origin (Table 3.4), while the bubble size indicates the average GRDP year 2011-2015 in Billion Rp*

Figure 3.4. Competitive position of each industrial origin in Cianjur Regency (2011-2015). Numbers in the bubbles refer to the sectors in Table 3.4.

c. Bekasi City

LQ and P values of Bekasi City are shown in Table 3.5. Eleven sectors are identified as basic sectors with electricity and gas activities having the highest score (3.87). The transportation and storage sector ranked in second position with 2.05. From the P values which were in the range of +/- 0.09, it can be concluded that all sectors in Bekasi City were growing and declining insignificantly, except for construction which is increasing in agglomeration growth with a P value of 0.19. In this region, the food-related sector is classified as non-basic sector with very low LQ value of 0.08, while water-related sector is categorized as basic sector with LQ value 1.05.

Table 3.5. LQ & P Values of Bekasi City Year 2011-2015

No.	Industrial Origin	AVG GRDP*	AVG LQ	Category	P
1	Agriculture, forestry, and fishery	325	0.08	Non Basic	**-0.09**
2	Mining and quarrying	-	0.00	Non Basic	-
3	Manufacturing	18,364	0.85	Non Basic	**-0.08**
4	Electricity and gas	1,020	3.87	Basic	0.02
5	Water supply, sewerage & waste	41	1.05	Basic	**-0.01**
6	Construction	4,688	1.18	Basic	0.19
7	Wholesale and retail trade	11,731	1.47	Basic	**-0.03**
8	Transportation and storage	4,543	2.05	Basic	**-0.05**
9	Accommodation & foods services	1,733	1.44	Basic	0.07
10	Information and communication	1,018	0.69	Non Basic	0.03
11	Financial intermediary services	1,333	1.14	Basic	**-0.03**

No.	Industrial Origin	AVG GRDP*	AVG LQ	Category	P
12	Real estate activities	845	1.48	Basic	0.03
13	Business activities	208	1.06	Basic	0.03
14	Administration & defence, social	1,031	0.95	Non Basic	0.00
15	Education	955	0.79	Non Basic	**-0.03**
16	Human health & social work	519	1.61	Basic	**-0.09**
17	Other service activities	1,366	1.46	Basic	**-0.06**
	Total	49,718			

Source: own analysis (*billion Rupiahs)

Agriculture sector is in the same cluster as manufacturing sector in Bekasi (see Figure 3.5). Both of them are located in the disadvantaged cluster (quadrant IV) during the years 2011-2015. Quadrant I is occupied by 5 sectors, while the remaining clusters i.e. quadrant II and III are filled by 1 sector and 5 sectors respectively.

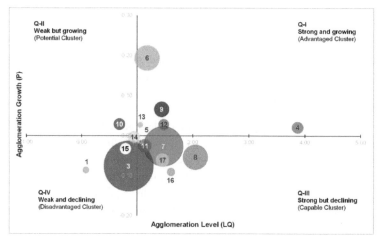

Notes: Numbers in/outside the bubble represent sectors or industrial origin (Table 3.5), while the bubble size indicates the average GRDP year 2011-2015 in Billion Rp

Figure 3.5. Competitive position of each industrial origin in Bekasi Regency (2011-2015). Numbers in the bubbles refer to those in Table 3.5.

3.4.2 Analysis of water, energy, and food-related sectors

The water-related sector is defined not only as domestic water supply, but also sewerage, waste management, & remediation products and services, but excludes agriculture or irrigation water. As depicted in Figure 3.6, the characteristics of water-, energy-, and food-sectors in each local government are shown. The water-related sector is not basic in Cianjur and Karawang (LQ 0.41 and 0.78 respectively). The value of the water-sectors GRDP is significantly smaller compared to the food and energy related sectors in those two areas. However, a different result is shown in Bekasi during 2011-2015, where the water-related sector is a basic sector with LQ 1.05. The water-related sector in Cianjur and Karawang was growing slightly as well as being in the 'advantaged' cluster. On the contrary, in Bekasi, the water sector is declining both in growth and the advantage of the cluster. The highest LQ score in energy-related activities was in Bekasi with 3.87, followed by Karawang and Cianjur with 1.57 and 0.17 successively. This sector is growing but not significantly in all the three regions. The food-related sector is divided into three sub-sectors; agriculture, livestock, hunting, & agriculture services, forestry & logging, and fishing. GRDP values of the food-related sector in Cianjur and Karawang during 2011-2015 were relatively high. However, the food-related sector in Cianjur is a basic sector with steady growth, while in Karawang it is classified as non-basic sector and the growth is slowing. Additionally, as a city, Bekasi is no longer a food producer. The LQ value of food-related sector was 0.08, compare to Karawang (0.46) and Cianjur (4.3).

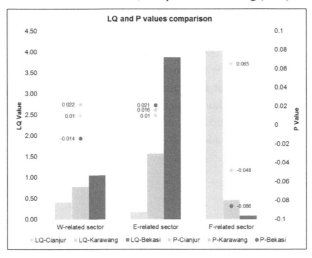

Figure 3.6. Comparison of LQ (bars) and P Values (dots) of WEF-related sectors in the three study regions. Dashed blue line indicates LQ = 1

Cianjur as an agriculture-based region with more than 265,000 hectares agriculture area shows its robustness in agriculture, forestry and fishing sectors. Additionally, with only 13 large manufacturing units, a population density of 621 inhabitants/km, and 0.4% population growth, the water and energy related sectors are less dominant compared to two other regions. Similar congruence can be seen in Karawang Regency, where industrialisation has shifted the sectors, especially food-related sector. The number of large industry units in this region is considerable (954 unit), with additional intermediate and small industry approximately 9,290 units. In Bekasi City, the labour force in agriculture sector is only 1%, while the biggest portion (74%) is in manufacturing, trade and services sectors. This situation is in line with the LQ score of Bekasi in the three sectors. Overall, these results indicate that WEF-related sector are growing slightly except

43

the water and food-related sectors in Bekasi and the food-related sector in Karawang which were declining during that period (2011-2015).

As shown in Figure 3.7, the characteristics of each sector (WEF) in the three regions are more clearly indicated in the competitive position chart. The energy-related sector in Karawang and Bekasi and the food-related sector in Cianjur are positioned in the advantaged cluster. The situation of the food-related is particularly interesting in Karawang where 96,482 ha of paddy field and 16% of the labour force constitute this sector, which are considerably higher than is Bekasi. However, as can be seen in Figure 3.7, this sector was located in quadrant IV in Karawang as well as in Bekasi which has only 475 ha of agricultural land and 1% agriculture-related labour force. The water-related sector is not located in quadrant I in any region, however the role of the water-related sector is very significant in supporting especially food-related and energy-related activities in all three regions. Agricultural water supply is excluded in the definition of water supply in these statistical data, mainly because there is no water pricing scheme in the regions, so its added-value is extremely hard to quantify in such an analysis. This is unfortunate, and it is likely that the water-related sector would position in a different quadrant in all regions if agricultural water usage as accounted for. Additionally, water is relatively difficult to be quantified in terms of GDP/GRDP calculations, since its role is generally as an 'indirect' value added.

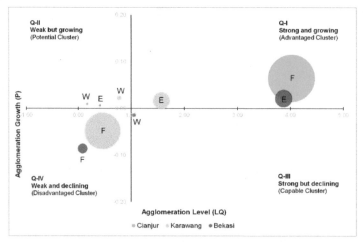

Figure 3.7. Competitive position of WEF-related sectors based on LQ and P values

3.4.3 Trends of WEF-related sub-sectors in Karawang Regency

Because of its special characteristics in the development goals that combines both agricultural and industrial activities, Karawang Regency has been analysed up to the sub-sector level. The agglomeration and trends of water, energy, and food-related sub-sectors in Karawang Regency were analysed using GRDP data of Karawang Regency at 2000 constant market prices (2000-2010) and at 2010 constant market prices (2011-2015), and compared with the same data and period of West Java Province as a benchmark. Figure 3.8 shows the comparison of LQ values in the three sectors of GRDP, where between 2000 and 2015, the water-related sector was categorized as a non-basic sector with LQ values ranging from 0.2 to 0.8. However, there was a significant increase between 2010 and 2011. There is a notable difference in the energy-related sector, where it is classified as basic sector from 2004. The most likely causes of this change in situation is the recent production of oil and gas both onshore and offshore. On the other hand, a clear decreasing trend can be seen in food-related sector. Starting from 2001, the value of LQ of this sector decreased gradually, and is now classified as a non-basic sector.

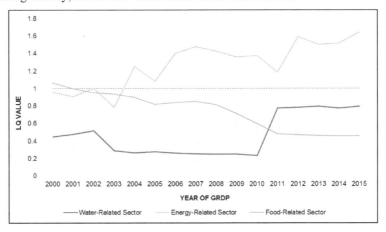

Figure 3.8. Agglomeration Trends of WEF-related sectors of Karawang Regency (2000-2015)

When broken down, the trends of each sub-sector from 2000 to 2015 are illustrated in Figure 3.9. The food crops sub-sector (green line, Figure 3.9) was basic before 2005, but started declining thereafter. In general, the decline of sub-sector concentration may be a result of agricultural land conversion and the influence of other agricultural based-regions that produce more agricultural commodities. A similar situation is seen in the fishery sub-sector in Karawang.

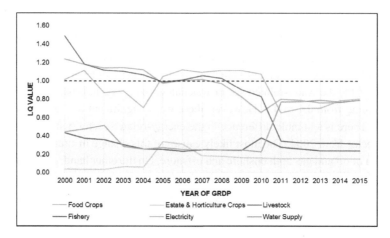

Figure 3.9. Agglomeration trends of WEF-related sub-sectors in Karawang Regency (2000-2015)

The shifting of the competitive position of WEF-related sub-sectors in Karawang can be clearly seen in Figure 3.10, where the green and yellow bubbles indicate the competitive position of each sub-sector in year 2001 and 2015 respectively. The movements from one cluster to another are shown for food crops, fishery, and electricity sub-sectors. The agglomeration growth of those sectors was relatively low, but the agglomeration level were significantly changed (i.e. a significant shift between quadrants is observed). On the other hand, the remaining sub-sectors such as estate & horticulture crops, livestock, and water supply also moved, but remained within the same cluster or did not move far into other clusters.

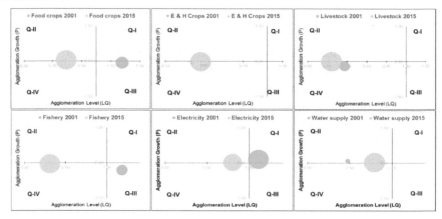

Figure 3.10. The trends of competitive positions of WEF sub-sectors in the years 2001 and 2015 in Karawang

Figure 3.11 shows the competitive position of WEF-related sub-sectors in Karawang during the period 2000 to 2015. Three of four sub-sectors were located in the disadvantaged cluster (food crops, livestock, fishery, and electricity sub-sectors), while the rest were in the potential cluster.

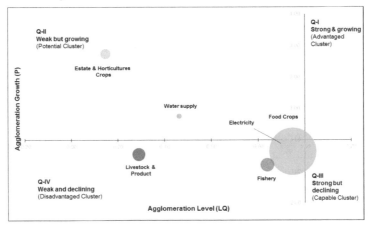

Figure 3.11. Competitive position of WEF sub-sectors in Karawang Regency (2000-2015)

It is interesting to note that in all three study sites, the agglomeration growth of WEF-related sectors was slower compared to the same sector at the provincial level as the benchmark, and they are not expected to become a basic sector in the future. The food-related sector was basic in the agriculture-based region (Cianjur), while energy-related and water-related sectors were non-basic. Additionally, water-related and energy-related sectors were basic in Trade, Service & Manufacture-based Region (Bekasi), while food-related sector was non-basic. Three sectors were situated in the advantaged cluster, while the rest were mostly in the potential and disadvantaged cluster. None of the water-related sector in the three regions is located in Quadrant I. The results show that water and food-related sectors were not basic, growing slower, and not expected to be a basic sector in the future.

3.4.4 Proposed strategies for WEF-related sectors and sub-sectors in Karawang Regency

There is a need for local stakeholders and policy developers to better understand the WEF clusters and their recent development trends in order to decide upon appropriate regional economic growth and development policy and to draw up suitable strategies and programs to assist sectors and firms within the cluster (Blakely and Bradshaw, 2002). Pearce and Robinson (2001) proposed a model of so-called grand strategy clusters within the four quadrants, where sectors or businesses are categorized based on the competitive position and growth. A set of expectant options can then be considered for determining the most proper grand strategies in order to facilitate growth or managed decline or certain sectors. The strategies can be used by government institutions and business or enterprises in the water, energy, and food-related sectors. The proposed strategies are expected to bring alternative options in the form of improvement activities, promotion to a better position, or even reduction and liquidation after considering local and national budget, targets, and planning.

After identifying the agglomeration of each sector, and putting them together in competitive position charts, the next stages are to analyse clusters formed in the charts and to determine proper strategies for each industry or sector. From the competitive position analysis (Figure 3.7 and Figure 3.11), there are three sectors positioned in quadrant I (advantaged cluster), while three sectors and two sub-sectors are in quadrant II (potential cluster). Additionally, one sector is located in quadrant III (capable cluster), while two sectors and four sub-sectors are in quadrant IV (disadvantaged cluster) (see Table 3.6). Some strategic options can be taken into consideration by stakeholders both government institution, private sector, and society related to water, energy, and food sector in Karawang Regency.

Table 3.6. Possible strategies for WEF-related sectors in Karawang Regency

Position	Sectors/sub-Sectors	Strategies	Example of practical implementation
Quadrant I (advantaged cluster)	• energy-related sectors	• concentrated growth • vertical integration • concentric diversification • innovation	• market expansion, product and service innovation • firm acquisition by regional owned enterprises • allocating sufficient budget to develop basic source of the sector
Quadrant II (potential cluster)	• water-related sectors • estate & horticulture crops • water supply sub-sectors	• reformulation of concentric growth • horizontal integration • divestiture • liquidation	• market expansion, product and service innovation • firm acquisition by regional owned enterprises • public-Private Partnership (PPP) scheme in providing drinking water • increasing the use of potable water
Quadrant III (capable cluster)		• concentric diversification • conglomerate diversification • joint venture	
Quadrant IV (disadvantaged cluster)	• food-related sector • food crops sub-sector	• turnaround or retrenchment • concentric diversification	• changing the development target from agriculture to industrial-based

Position	Sectors/sub-Sectors	Strategies	Example of practical implementation
	▪ electricity sub-sector ▪ livestock sub-sector ▪ fishery sub-sector	▪ conglomerate diversification ▪ divestiture ▪ liquidation	▪ food production diversification (ex. from rice to potato and soya bean or livestock and fishery production)

The basic principle of selecting strategies for sectors in quadrant I as the 'advantaged' position is to maintain the competitiveness, concentration, and growth with certain innovation and new developments within the value chain. For firms and sectors within this quadrant, a concentrated growth approach can be carried out by increasing and promoting market development, product development or both. Another possibility for the sectors in quadrant I is to undertake vertical integration option, in case its resources have surpassed the total demands. Vertical integration is an action to acquire other firms either a firm in very early stage of the value chain (backward integration) or a firm in a later part of the value chain (forward integration). To mitigate potential risks related to the product and service limitations, concentric diversification can be considered by investing certain budget and other capital in its proven basic fields. It is basically an approach to create a new portfolio through the acquisition of an established business or sector.

The example of sector in this quadrant I is energy-related sector. One of the reasons that makes the energy-related sector in Karawang Regency positioned in the advantaged cluster is because this region produces both oil and gas. Thus the level of concentration and competitiveness of this sector is quite strong. However, in the electricity sub-sector, its position is still in the disadvantaged cluster. Some possible strategies that can be conducted are increasing the local potential of electricity production, especially renewable energy such as solar energy instead of fossil fuel-based electricity. Solar energy can be utilized both in urban and rural areas to support electricity supply for domestic, industry, and agricultural sectors. In addition, the government can support it by putting a sufficient budget and subsidies to develop basic sources of the sector and networks or markets. Those efforts will certainly become one of the supporting actions for the achievement of Central Government's target for solar energy of 238% in 2019, and electrification ratio up to 100% in 2019. A drive to solar electric generation would help meet national targets in a sustainable way, without seriously compromising other natural resources.

Identifying the non-competitive factors and conducting improvement of the sector's inputs are the basic approaches that must be considered for sectors located in quadrant II (potential cluster). There are four strategies that can be carried out: reformulation of concentrated growth; horizontal integration; divestiture; and liquidation (Pearce and Robinson, 2001). Reformulation of concentric growth is the best option to choose for sectors able to achieve targets in a relatively growing market. Nevertheless, if sectors are relatively less competitive, the second option of horizontal integration is also possible. Principally, this strategy seeks to acquire other firms/sectors in the same level and field to make the competition less severe and increase growth by improving market and product development. One undesirable option is divestiture, where the authority has to sell off subsidiary or entire investments as an ongoing business that may cause inability to fully control all the processes within the sector. In the government businesses' perspective, sometimes this kind of policy is provoking criticism from the public due to possible uncertainties in providing public products and services. Finally, liquidation is the other options to be taken into account. It is slightly different with the divestiture, wherein this action, the ongoing business is not intended to be further operated. This option is reasonable to avoid further detriment of the firm or sector in the long run.

Water-related sector and water supply sub-sector were the two industries positioned in quadrant II. In the sub-sector of water supply (potential cluster) for instance, government and private sector can improve the competitiveness and market of this sub-sector through cooperation using public-private partnership (PPP) schemes in the provision of drinking water and raw water. Improvements and innovations of the products and services can also be done by government and private companies to attract users to the products. This could include increasing potable water and bottled water supplies, increasing the pipe water network coverage, improving system performance and reliability, etc. Another option is to acquire private bottled water companies by government owned enterprises to ensure the availability, affordability, and quality of the products. Increasing water production from local resources such as ponds, artificial lakes, rainwater harvesting, and other water conservation programs will also support this sector. Those strategies are highly relevant to the central government's target to improve the access to clean water from 65% in 2014 to 100% in 2019. Improving system efficiency, and trying to move away from bottled supplies (which are more costly and less sustainable due to elevated energy and water requirements in the production process) for potable purposes would boost sustainable development initiatives in the region, and could free up water resources that are currently lost in inefficient networks for use in the expanding industrial sector and could also be used in a growing agricultural sector should this develop according to regional plans.

For the sectors and firms in quadrant III (capable cluster), observing the market and consumers and then minimizing the investment based on that evaluation are plausible options to be taken. Concentric diversification can be applied to the business or firms by encouraging innovation both in markets and products. Another choice is to carry out conglomerate diversification, where the firms diversify the portfolio of their investment and financial risk. The last option for this cluster is a joint venture which focuses more on creating cooperative agreement with other domestic or multinational firms. The main purpose is to improve the performance of each part within the two or more firms collaboratively to achieve the targets, and thereby to raise the performance of the sector as a whole. According to the current results, there were no single sector located in this capable cluster (quadrant III).

Two basic approaches that can be considered for the sectors or businesses positioned within quadrant IV (disadvantaged cluster) are preparing for transition or trying to manage the decline. Turnaround or retrenchment is the option to change the business direction starting from increasing efficiency and cost/assets reduction. It can be proceeded to the next phase by diverting the resources through investment in other sectors. Either concentric diversification (market and product) or conglomerate diversification (business and financial portfolio) is involved at this stage. The final strategies that need to be noticed in this quadrant IV firms are divestiture and liquidation.

Things are slightly different in the food-related sector, where almost all of its sub-sectors are in disadvantaged clusters except estate & horticulture crops. Excessive industrial development may cause agricultural land conversion either directly or indirectly. Additionally, climatic factors and the availability of water, especially in the dry season causes a considerable amount of food crop production to be not optimal. One of the strategies that can be considered is to change drastically the direction of the regional planning from agricultural-based to industrial-based development. The strategic position of Karawang means that it has large potential to develop industrial, trading and service sectors, but at the expense of the agricultural sector. However, this strategy is counter to the regulation of sustainable agricultural land, where rice fields in Karawang regency must be protected from land conversion. Driving industrial and service growth would also likely stress the demand for water and energy, and it is not clear if it could be accommodated in a sustainable manner without compromising local natural resources. Another possible strategy is the diversification of agricultural products, including livestock and fisheries. By diversifying both the product and the market or users (concentratic diversification) and its business and financing strategies (conglomerate diversification) are expected to increase the competitiveness of food-related sectors in this region. A further strategic option is a combination of industry and agriculture, where the industrial sector to be developed will be prioritized to the food processing industries that can absorb the existing agricultural products and increase its value added. The central government's target in food-related sector in 2019 are rice production (26%), soya bean (109%), beef (67%), and fish (51%) among others. A diversified strategy with financial support from local and central government are expected to support the achievement of the targets.

3.4.5 Sustainability and nexus perspective in WEF resource management

This study provides a first assessment of the current levels of growth and agglomeration in WEF sectors in Indonesia, giving policy makers a first idea of recent trends and what could be reasonably expected in the next few years. In terms of sustainable development, Section 4.4 gives some indication of how the WEF sectors could be developed in a sustainable way and in accordance with national targets. However, current trends appear to show a shift away from targets, and may not be sustainable in the long term if not managed properly. In addition, it is becoming apparent that the WEF sectors can no long be considered in isolation, and that changes in one sector is likely to have implications for the other sectors (Kenway *et al*. 2011; Bizikova *et al*. 2013; FAO 2014; El Gafy *et al*. 2016). The methods used in this study prohibit such a 'nexus' approach, although such an approach is becoming increasingly necessary. While these results give a first order impression of recent and potential short-term changes to the WEF sectors in the three study areas, further analysis would be required in order to conduct a full 'nexus assessment' accounting for cross-sectoral impacts of economic growth and development. That being said, there are some reasonable hypothesis that can be put forward. For example, a lack of water storage has caused flooding in some areas during the rainy season, and conversely in the dry season, crop failure occurs due to lack of water supply and sufficient storage. Another pressing issue is water quality. Effluents from industrial, agricultural and domestic activities, some of which are shown to be growing rapidly, has polluted almost all segments of rivers in this area. Policy is required to establish new artificial lakes/ponds, infiltration wells, bio-pore holes and promoting rain harvesting. The expected benefits are increasing storage in the wet season, mitigating flooding and also providing extra water in the dry season to maintain the growing industrial sectors. Additional water could also aid agricultural expansion, in line with policy targets. In addition, a drive to solar energy is likely to reduce regional water demand in the energy sector while productivity can be maintained. This leaves additional water resources for other activities whereas it may be presently consumed for energy production. As a result, food production could be positively impacted, even in water-scarce seasons, further contributing to policy targets.

At the same time, energy is required to support food production activities (e.g. harvesting, fertilizing, pumping, lighting, processing, and transporting) and water-related production. The notion to develop urban and rural solar energy in this area is strongly relevant in the context of its poor status of air pollution. Energy interventions propose solar urban and rural planning to improve the utilization of solar energy and examine its implication to water, energy, and food security. Apart from its expected contribution to the national target of solar energy application (increasing 238% in 2019 from base year 2014), this policy is intended to satisfy local energy demand for any purpose, and to reduce the use of fossil energy that cause air pollution problems. The use of solar energy is expected to give positive impact in combination with artificial lakes/ponds on water needs for food production, particularly rain fed agricultural area in the dry season. It may contribute to avoiding crop failure due to droughts and support livestock water needs. Solar energy applications in an urban context could include use in telecommunications, service and trading, building equipment, public lighting, and industrial use, while in rural areas it can be used for applications including water pumping, crop drying process, greenhouse heating, livestock ventilation, aeration pumps in fishery, fish or shrimp cultivation, insect pest trap, and solar sprinkler. This implementation should be supported by government through subsidy or incentive schemes, in order to stimulate other stakeholder e.g. private sectors, educational institution, and community to succeed this solar energy policy. By evaluating the WEF sectors through their value-added in GRDP, decision makers are better equipped to think, plan, and implement the appropriate actions to ensure the availability of those resources. By subsequently adopting a nexus approach, it is possible the sustainable development in the region could be attained.

GRDP data are essentially arranged based on a production and expenditure approach. In this study, production-based GRDP by industrial origin denotes the basic measure of value-added emerging from various kind of economic activities and production in a region. Basic production factors are implicit in GRDP and cannot be easily separated from it as they are not directly reported in GRDP statistics. Therefore, the analysis conducted in the chapter implicitly accounts for basic production factors, but is not able to explicitly account for the impact of changes in various production factors to change in GDRP.

Despite the limitations, these results could be used as preliminary information for decision makers to make the future planning and actions and also to review recent progress towards existing goals. They can also be used to put development in a wider sustainability/nexus context.

3.5 CONCLUSION

Strategies for each cluster have been identified based on the analysis of agglomeration level and growth and then putting all the sectors in the four-quadrant matrix. Local stakeholders both government, private sectors, and community should pay attention to these particular sectors to ensure the availability of the commodities. This preliminary evaluation will give a better understanding and more comprehensive insights. All the strategies have to be further analysed to be properly implemented in the decision making process by considering local and national budget, targets, and planning.

All the definition of water, energy, and food-related sectors are based on the definition from statistics agency which may slightly different with other definition. One of limitations is the exclusion of agricultural or irrigation water in the definition of water supply. The accuracy of the results in LQ computation will strongly depend on the accuracy of the dataset. To avoid possible seasonal and annual bias of dataset in the calculation, we used the average of more than 5 year data series. In this study, the data used were basically the statistical data series of 16 years of GRDP (2000-2015) that has been verified and analysed by the national statistical agency. Unfortunately, the Agency does not report on potential sources of error or the errors themselves. Therefore it is very difficult to assess the errors involved as they are not reported. In principle, LQ and CP methods are not the tools that need to be validated statistically using certain sample size to analyse the uncertainties and errors. However, there is also an option to do such statistical analysis, for instance dartboard test for LQ established by Guimarães et al (2009). Further study and analysis using the direct method such as census and its combination are also recommended to be conducted.

To conclude, this composite method is useful and is expected to assist local stakeholders in evaluating preliminarily the general situation of the basic and non-basic sectors and also the dynamics of the WEF-related sector in their economic development. The relevance of economic activities development characteristic and its growth is clearly supported by the current findings.

4

GROUP MODEL BUILDING ON QUALITATIVE WEF SECURITY NEXUS DYNAMICS

Abstract

This chapter develops a qualitative causal model of a water, energy, and food (WEF) security nexus system to be used in analysing the interlinkages among those and other sectors that influence and are influenced by each other in a local context. Local stakeholder engagement through a group model building (GMB) approach was applied in Karawang Regency, Indonesia, to develop the model with the goals of improving problem understanding, raising consensus among participants, and building acceptance and commitment regarding the subsequent development of a quantitative nexus model. After recognizing the issues regarding WEF sectors in the study area and eliciting opinions about nexus interactions, the next stage was to build a conceptual framework to describe the nexus system and to develop an integrated causal loop diagram (CLD) that describes critical system (inter-)linkages. The developed Karawang WEF security (K-WEFS) model is composed of six sub-models with water, energy and food sectors as endogenous factors. In addition, population, economic and ecosystem services were considered as exogenous drivers of the system. It is expected that all the major internal and external factors and drivers are covered, including possible feedback mechanisms, and key variables will be analysed further in the system. The future achievement of WEF security targets can be based on robust evaluation and planning processes underpinned by thorough understanding of whole system dynamics and the impacts of changes in the linked sectors, even in a qualitative way. In this way, a first step towards breaking silo thinking in regional planning may be attained.

Keywords: group model building, causal loop diagram, WEF security, nexus modelling

This chapter is based on:

Purwanto A., Sušnik J., Suryadi F.X., de Fraiture C., (2019), *Using group model building to develop a causal loop mapping of the water-energy-food security nexus in Karawang Regency, Indonesia*, Journal of Cleaner Production 240, Elsevier, DOI: https ://doi.org/10.1016/j.jclepro.2019.118170 (Published)

4.1 INTRODUCTION

Water, energy, and food (WEF) security are a major topic that is increasingly discussed globally. The interaction among their components internally and the interconnection with environmental conditions, social, governance, and political situations, makes this issue immensely complex. Achieving a certain level of WEF security simultaneously is a complex challenge that will influence, and is influenced by other sectors including social, political, and environmental conditions (Bizikova *et al.*, 2013; Endo *et al.*, 2015). Resolving one problem partially without considering its interlinkage could shift problems from one resource perspective to another and may cause unexpected effects *(Kenway et al.*, 2011; Bizikova *et al.*, 2013; FAO, 2014; El Gafy *et al.*, 2016). Focusing only on certain aspect of security, without considering others may cause unbalanced supply or resource use, and ineffective target achievement. In addition, 'optimising' across the nexus may mean admitting that some targets in certain sectors cannot be fully met, but on average across all sectors, more targets are achieved more satisfactorily.

The basic concept of WEF security nexus was developed and extensively discussed at the Bonn 2011 Conference. In its background paper, the nexus approach defined 'an approach that integrates management and governance across sectors and scales' (Hoff, 2011). Nevertheless, there are still knowledge gaps in this approach including analytical framework disharmony for overcoming institutional disjunctions and power imparity among sectors. Furthermore, there is no sole technique that is able to be applied for every specific circumstance suitably (Endo *et al.*, 2015). Thus, to deal with different and specific situations in different regions, deconstruction of the nexus approach (Lele *et al.*, 2013) and specific context elaboration (El Gafy *et al.*, 2016) have to be considered in order to make more effective and contextualized solutions on water, energy, and food security, and to assist decision makers in managing resources in a meaningful way. Thus, local stakeholder involvement in combination with simple and understandable analysis tools are needed in doing evaluation, planning and decision making process of WEF security in a region.

Several models, tools and approaches have been developed to analyse the interaction between water, energy, food security and the environment (IRENA, 2015; Daher and Mohtar, 2015; El Gafy *et al.*, 2016; Martinez *et al.* 2018). Those tools can be conceptual (qualitative), quantitative, and some combination of both qualitative and quantitative. One major system thinking approach is system dynamics, established by Jay Forrester in 1961, some concepts of which are employed here for its flexibility and robustness in portraying a complex system to non-experts. It is a method that typically used to elaborate a complicated system, multi-feedback-loop, multi-disciplinary and non-linear system that contains of three main components; stocks, flows, and converters (Sušnik *et al.*, 2013). The competence to identify and trace behaviour pattern in a system, instead of focusing only on individual situations, is needed in dynamic thinking (Simonovic, 2009). Meadow

(2009) identifies that a system comprises three things i.e. elements, interconnections, and function or purpose. It can be formed as part of a qualitative and quantitative approach.

Numerous studies have attempted to explain the interconnectedness among water, energy and food sectors and variables through quantitative approaches. Martinez *et al.* (2018) applied a mixed method of fuzzy cognitive maps (FCM) to understand WEF nexus interconnections and their relative strength using expert input. Graph theory was used as a basis to develop the analysis which highlights how different experts' views can be captured in such a fuzzy mapping approach. One downside however is that complex feedback relations may be more difficult to capture using FCM. Some other scholars such as Ma *et al.* (2019) and Liang *et al.* (2019) study the nexus in energy, environment and economy by employing decomposition and decoupling methods. The methods successfully quantified carbon dioxide intensity due to economic development in China. Furthermore, (Zhang *et al.*, 2018) investigated few more methods that have been applied by researcher to examine quantitatively the nexus of water, energy and food such as mathematical statistics, computable equilibrium model, and econometrics among others.

However, in some cases, extensive data may not be available in order to develop a quantitative model, or the main objective may be simply to gain insight into the behaviour and structure of a complex system such as the WEF nexus. One simple but effective qualitative modelling approach not requiring extensive data or expensive software that is being applied by many scholars analysing complex systems with stakeholders is that of causal loop diagrams (CLD). CLD development is very useful in assisting system dynamic modellers and non-expert stakeholders in capturing the complicated interactions among variables qualitatively and can form a basis for further system analysis (Haraldson, 2000; Mirchi *et al.*, 2012; Binder *et al.*, 2004; Inam *et al.*, 2015). CLDs offer non-experts an easy and intuitive way in which to view entire systems, and critically the linkages between system elements (i.e. the nexus), including feedback processes. As no modelling experience is required, CLDs can be very attractive to engage stakeholders in the process of developing quantitative complex systems models. A number of researchers have employed CLDs in analysing diverse complex systems (Mirchi *et al.*, 2012; Sušnik *et al.*, 2012; Bala *et al.*, 2014; Ghashghaie *et al.*, 2014; Inam *et al.*, 2015). Furthermore, one of the participatory approaches in developing CLDs is the group model building (GMB) method (Vennix *et al.*, 1996). Numerous studies have explained the usefulness of participatory and GMB in constructing CLDs (for instance Vennix *et al.*, 1996; Luna-Reyes *et al.*, 2006; Winz *et al.*, 2009; Goh *et al.*, 2012; Hovmand *et al.*, 2012; Inam *et al.*, 2015; Kotir *et al.*, 2017; Rich *et al.*, 2018).

The above section clearly shows a research gap – the need to better implement non-computational approaches to better understand the WEF security nexus in a robust way that is to be taken seriously by local planners and that can be understood by them. Such an approach may be extremely valuable either in data scarce settings, or as a first stage towards quantitative model development as a way to better understand system dynamics. Therefore, the main objective of this work is to develop a qualitative causal loop diagram (map) of the water-energy-food security nexus in Karawang Regency, Indonesia, with the aim of elucidating to local stakeholders the complexity of the system without recourse to a complicated modelling and data collection exercise. Such an exercise can highlight the complexity of such systems to stakeholders, and by thinking across traditional departments, can also help to break silo-thinking, moving towards systems thinking. This chapter describes the development of a qualitative causal model of water, energy and food security in the local context of the Karawang Regency, Indonesia, to be used in analysing the interlinkages among those three sectors and other sectors that influence and be influenced by each other. Group model building as part of stakeholder's engagement was applied in the study area to improve problem understanding, raise consensus and build the spirit and commitment of related stakeholders. The work aims to uncover unexpected high-order effects of policy objectives, and could lead to policy makers exploiting synergies and avoiding or mitigating detrimental impacts across the nexus. The rest of this chapter is organised as follows: Section 4.2 outlines the main methodological approach used in the study, namely causal loop diagrams and group model building. Section 4.3 describes the case study, and presents the major results from the study. In addition, a detailed application of the developed qualitative model is described, illustrating the utility of this approach in practice. Conclusions and References complete the chapter.

4.2 METHODS

4.2.1 Causal loop diagram (CLD)

CLD is a qualitative approach that can be applied in the process towards developing quantitative system dynamic models (Binder *et al.*, 2004). CLDs are very helpful in assisting non-expert stakeholders in developing a better understanding of the main interconnections in a complex system, shedding light on critical feedback loops and connections that may otherwise not have been apparent. They also may therefore lead to a better understanding of the system behaviour, without any quantitative model ever being built. They therefore contribute to breaking traditional silo-thinking. Wolstenholme (1999) explains that CLDs are able to be applied as a stand-alone system and not necessarily need to be supported by computer simulation in developing subject and solving selected problems. Causal effects among variables in CLDs are connected by arrows with polarity either positive (+) or negative (-) to indicate their interdependency (Sterman, 2000). Connectors or arrows in the diagram function to deliver information from one variable (A) in the system to other (B). The arrows with a "+" sign show that change (increasing/decreasing) in A causes change (increasing/decreasing) in B in the *same* direction. Meanwhile, the arrows with"-" signs indicate an *opposite* direction from A to B (i.e. if A goes up, B goes down). System dynamics behaviour can be in a circular form of feedback loops, and self-reinforcing behaviour (positive feedback loop) and self-balancing behaviour (negative feedback loop) can result in the system. Normally, a positive feedback loop represents the continuity of growth or slowdown, while a negative one consist of causal links that try to fill the gap between desired and current condition (Mirchi et al., 2012). Another important notation in CLDs is delay. Indolence, fluctuation, short and long run effect of policies in the system can be caused by delays (Sterman, 2000). Table 4.1 summarizes visual representation of CLD's notation. Vensim modelling software, developed by Ventana System Inc., has been used to develop the CLDs in this study based on the GMB exercise. Therefore, all modelling languages, notations and formulas will be based on the Vensim system.

Table 4.1. Elements in causal loop diagrams

Notation	Description	Example
A $\xrightarrow{+}$ B Connector	change in A, causes change in B in the same direction. If A increases/decreases, B also increases/decreases	Temperature $\xrightarrow{+}$ Evaporation Cultivated land $\xrightarrow{+}$ Water demand
A $\xrightarrow{-}$ B Connector	change in A, causes change in B in the opposite direction. If A increases/decreases, B also increases/decreases	Infiltration $\xrightarrow{-}$ Run-off Groundwater table $\xrightarrow{-}$ Pumping cost
or R (+)	Reinforcing or positive feedback loop, if it contains an even number of negative causal links	Birth Rate R Population
or B (-)	Balancing or negative feed back loop, if it contains an odd number of negative causal links	Population B Death Rate
or $\xrightarrow{+}$ $\xrightarrow{-}$	Delay, the situation when the systems respond slowly in certain condition	Number of Plant Growing Harvest Rate

Source: Modified from (Mirchi et al., 2012)

4.2.2 Group model building

GMB involves several participants from various related parties, facilitated by a modeller/facilitator to build an expected model. Participants should have expertise relevant to different parts of the system under study (in this case, different WEF sectors in Karawang). Participants may or may not also be modelling experts. The idea is to build a conceptual map of the system using participant input to elucidate key system connections, interdependencies and feedback, giving a whole-systems view of the system under study, instead of a sectoral view. The conceptual map may then be taken forward for quantitative model development if desired. GMB is important in improving better understanding of the problem, aligning perception and building strategies to potentially improve performance of the system (Vennix, 1996). System dynamics GMB is useful and effective in developing more comprehensive qualitative-based understanding as a result of the involvement of multi-disciplinary experienced groups of researchers, managers and decision makers during its development (Luna-Reyes *et al.*, 2006; Richardson and Andersen, 2010; Goh *et al.*, 2012; Inam *et al.*, 2015; Rich *et al.*, 2018).

In general, GMB begins with a general discussion between the modeller, the gatekeeper and potential participants to define the problem (Figure 4.1a). Several crucial topics to be discussed in this stage are defining the boundary of the system, the complexity of the problem, the long-term and short-term impacts, the type of model to be developed and its aims, and potential stakeholders to be involved. Preliminary models and/or direct interviews can be chosen before starting the session of GMB. Figure 4.1b describes the steps of causal loop diagram (CLD) development through problem analysis, cause and effect elaboration, and feedback loop identification. Vennix (1996) identifies three goals of GMB i.e. creating a team learning environment to improve the problem understanding, raising the consensus among participants, and building the spirit of acceptance and commitment to final decisions.

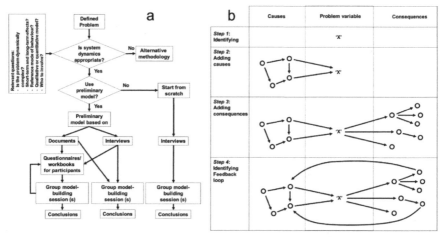

Notes: 'X' problem-variable; O other variables. Source: adapted from Vennix (1996)

Figure 4.1. The general stages in applying system dynamics GMB (a) and the process of causal loop diagram development (b)

4.3 RESULTS & DISCUSSION

4.3.1 State of the system

The Indonesian government has a strong commitment to achieving nationally set targets in water, energy, and food security as outlined in the national long-term and medium-term planning. However, several acute problems have obstructed the attainment of its national targets, i.e. resources mismanagement, lack of coordination, and authority imbalance among sectors, levels and scales (Bellfield *et al.*, 2016). The challenges are getting more complex in the current decentralization era where local governments (i.e. 34 provinces, 416 regencies, and 98 cities) have their specific local approaches and targets. A decentralized system can be an effective way to achieve national targets if each local government has a harmonious framework and implementation plan. Otherwise, the gaps among sectors and levels can get wider, and the targets are becoming more difficult to reach. Quincieu (2015) emphasizes the need of preferable and clearer roles, responsibilities, programmes and policies among district, provincial and central government in Indonesia. An integrated local framework on WEF security evaluation is importantly needed to build awareness and improve understanding among local stakeholders regarding the local WEF resources potential and their interlinkages.

As one of local regions in Indonesia, Karawang Regency plays an important role in supporting the achievement of national WEF security targets. This region is located downstream of the Jatiluhur reservoir, which is located within the Citarum Basin, Jawa Barat Province, Indonesia. Geographically, Karawang is situated between 107°02'– 107°40' East Longitude and 5°56'–6°34' South Latitude. Karawang Regency consists of 30 sub-districts, 297 villages and 12 special villages. It is one of the largest agricultural centres in Indonesia with rice as the main commodity. On the other hand, this region also focuses on industrial development as stated in its long-term (twenty-yearly) planning for 2005-2025. The total population of Karawang Regency in 2016 was 2,273,600, with a population density of 1,187 people km^2 and population growth in year 2015-2016 of 1.04% (Statistics Agency/BPS-Karawang 2016).

4.3.2 Basic concept of WEF security nexus in Karawang Regency

WEF security definitions used in this study refers to the availability, accessibility, and quality of those three sectors. In principle, resource availability is the existence of the resources physically at a certain minimum level to meet demand, while resource accessibility is defined as the ease of accessibility of the resources by people in a region at an affordable price. Furthermore in terms of quality, water quality defined as water that meets water quality standards locally. Energy quality means that the energy generated can be utilized continuously and safely regarding the impact to the environment. Three principles of nexus approach by Hoff (2011) namely investing to sustain ecosystem

61

services, creating more with less, accelerating access and integrating the poorest are employed in this research as the main concept in modelling qualitatively the existing condition of the study area.

The proposed local WEF security framework is depicted in Figure 4.2, and is applied in Karawang Regency, Indonesia. The grey boundary line reflects a regency boundary covering all the variables, links, and interlinkages within the region, while boxes that contain variables and links outside the boundary i.e. Jatiluhur Reservoir (JR), imported water (IW), imported energy (IE), and imported food/IF) are considered as transboundary or exogenous factors. The imported water, energy and food indicate that the main source of those commodities come from the outside Karawang Regency, and are not directly controlled or mediated by processes within Karawang. This basic concept seeks to describe internal interaction in WEF system (blue, yellow, and green boxes and arrows) and also interlinkages between the WEF system with other external driver factors i.e. population, economic development, and ecosystem services which is symbolized by grey boxes and arrows.

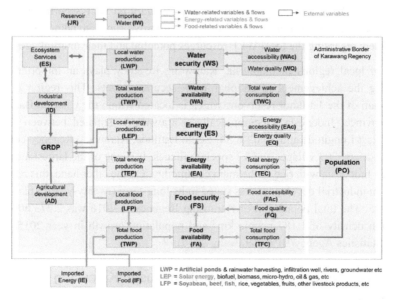

Figure 4.2. Basic concept of WEF security nexus in Karawang Regency, Indonesia

4.3.3 GMB workshop script

The one-day GMB workshop on WEF security nexus was conducted in October 2018 in the study area of Karawang Regency. It was preceded with a series of formal and informal meetings with all potential committees and participants in order to clarify the aims and objectives of the GMB exercise. The main session of GMB workshop focussed on the development of the integrated WEF security CLDs in the local context. Figure 4.3 describes the stages of the GMB workshop. The outcome of the workshop was expected to advance the understanding of all stakeholders to the problematic WEF security behaviour.

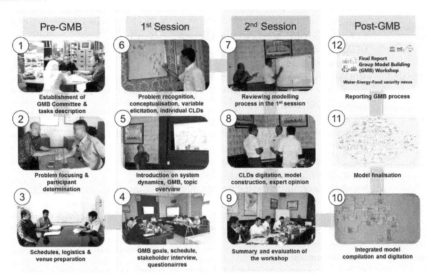

Figure 4.3. The stages of GMB workshop on WEF security nexus in Karawang Regency

Table 4.2 indicates several elements of script used in a system dynamics GMB exercise. Scripts are general patterns to facilitate participants in building a system dynamics model. Pre-GMB workshop activities were started approximately one month before the event, to ensure all the preparations regarding the committee, venue, materials and participants were ready to support all sessions of GMB workshop. Those activities brought a positive impact on increasing understanding of the topics, methods and both technical and non-technical matters that will be carried out together in the workshop.

Table 4.2. Elements of a typical GMB script

Field	Description
Description	Group model building (GMB) workshop on water, energy, and security nexus in local context involving related local stakeholders in these fields
Context	This script can be used in discussing important policies that need to be determined by involving several stakeholders
Participants	Agriculture Agency, Fishery Agency, Drinking Water Company, PJT II Public Corporation (reservoir authority), Public Work Agency, Environment Agency, Development Planning Agency, Singaperbangsa University of Karawang, local energy experts and Non-Government Organization (NGO).
Purpose	(1) to improve better understanding of the problem, (2) to align perception (problem framing), (3) to elucidate variables, and (4) to develop qualitative SDM in the form of CLDs

Field	Description
Primary nature of group task	• Presentation. Aligning the same perspective about the topic and methods used in the workshop • Divergent. Participants come from different institutions in local government level related to WEF sectors, more specifically from planning or research and development division.
Time	• Pre-GMB workshop (±1 month before the event) • GMB workshop (preparation ±30 minutes, main activities ±240 minutes, ±30 minutes evaluation)
Materials	Rounded rectangle/U-shaped table, desktop PC/laptop, projector, screen, sound recorder, white board, blank wall, sticky notes, moveable chair, board marker, camera/handycam
Input	State-of-the system, basic concept of WEF security
Outputs	Sub-causal loop diagrams, integrated causal loop diagram of WEF security nexus
Roles	• Modeller: listening to what being discussed and modelled during the session • Facilitator: organizing the workshop sessions • Gate keeper: initiating the project, identifying the participants, supporting the team • Recorder: documenting all the GMB workshop session either videos or photos
People in the room	Modeller, facilitator, gate keeper (optional/not all the session), recorder, participants
Steps	**Opening session** • The gate keeper delivered the opening speech to explain the overview of GMB workshop in general • Facilitator explained the workshop goals, schedules, methods and everything that need to know by the participants • Facilitator collected the open questionnaires from the participants that have been sent to participants beforehand • Facilitator opened the short session on stakeholder's interview **Introduction session** • Facilitator/modeller delivered the introduction session to explain about the research overview, system dynamics model (especially how to create CLD), and group model building approach. • Discussion, question-answer session moderated by facilitator/modeller. **Modelling session I** • Facilitator/modeller started to hands out the conceptual model, blank papers, markers and other materials to each participants. • Facilitator/modeller gave the chance to the participants to recognize problem, gave comment/feedback and conceptualized the initial model based on the preliminary conceptual model. • Participants included additional variables that need to be considered in the WEF model and elicits the reason/argument of each variables and put it in the sticky notes. • Facilitator/modeller asked the participants to create individual CLD in water, energy or food security in their blank papers. They walk around during the session to assist participants in doing their works. **Modelling session II** • Facilitator/modeller invited each participant in each sector to put their variables/CLD in the blank paper in the board. Other participants from the

Field	Description
	same sector were also invited to do the same thing and discuss it in front of the room.
	▪ They combined their individual variables and CLD under the supervision of facilitator/modeller.
	▪ Modeller digitised the sub-model CLDs in the SD software (Vensim PLE) and shows it to all the participants.
	▪ Facilitator/modeller repeated step 4a to 4c for two other sectors
	▪ All the sub-model CLDs were shown to all the participants and modeller starts to discuss on the integration of those three sub-model CLDs.
	▪ Preliminary WEF security model that has been made before by the modeller was used as general guidance in order to make the participants easy to understand. Some variables and arrows were changed based on the perception and opinion of all participants.
Evaluation criteria	▪ The perception and knowledge of participants to the topic and methods need to be identified by doing short interview, to make sure whether we need to take longer time for introduction session or not
	▪ All the participants have agreed about the effectiveness of group model building and qualitative dynamics model in analysing the policies and finding the best solution.
	▪ Open questionnaire helps participants in understanding the topic and helps facilitator in organizing the sessions

Source: the script is adapted from Hovmand et al. 2012

Ten institutions related to the topic participated actively in building the models. The water sector was represented by Drinking Water Company (PDAM), Irrigation and Reservoir Authority (Jasa Tirta II public corporation) and Public Work Agency. Agriculture and Fishery Agencies played a role in providing input regarding the food sector. A local energy expert, who is a former Energy and Mineral Agency employee, was personally involved because the authorities of the institution related to mining and energy had been taken over by the Province. The Development Planning Agency, Singaperbangsa University and an NGO member were also involved in the workshop to share information and feedback on both endogenous and exogenous variables in the WEF system. Differences in point of views and experiences among participants were presented very dynamically in the interactive discussions guided by the facilitator/modeller. Various information, ideas, input, and qualitative data were delivered by the participants in the workshop. A recorder played an important role in documenting all activities carried out in each session, in the form of photos and videos. After all stages have been carried out, all participants agreed on the formulation of three sub-models which were then incorporated into one integrated model so-called Karawang WEF security (K-WEFS) model, which is the main output from the GMB exercise.

4.3.4 WEF security sub-models

a. Water security sub-model

The developed water sector sub-model is shown in Figure 4.4. The source of raw water in this region comes from surface water (rivers, lakes and ponds) and underground water. Imported water as the main water source is interpreted as water supply from Jatiluhur Reservoir whose location is administratively located outside Karawang Regency. On the other hand, water demand comes from agricultural, domestic and industrial activities. The supply and demand of water affect water availability in this area. Problems related to water availability arise in this region during the rainy and dry season. The lack of water storage in the area leads to flooding in some areas during the rainy season. Conversely in the dry season, the crop failure frequently occurs also due to the lack of reliable water supply. In addition to the availability factors, water accessibility and particularly water quality were discussed intensively in the model preparation. This is due to the fact that this regency is not only an agricultural area but also an industrial area with potential pollution to the rivers and other water sources (Figure 4.4), which could impact negatively on agricultural production. Artificial ponds development is focusing on the green space in individual housing, industrial and rain fed agricultural areas in this region. Thus, the enlargement of those sectors' developments is expected to increase the number of artificial ponds. In the end, it could enhance local water production and conversely reduce the imported water demand.

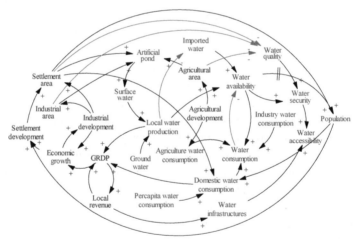

Figure 4.4. Water security sub-model

b. Energy security sub-models

Figure 4.5 shows the sub-model of energy security developed for Karawang Regency, which is mostly focused on electricity. The main source of electricity for Karawang

Regency comes from a hydropower plant located in Jatiluhur Reservoir (and therefore interpreted as imported energy). In addition, several sources with very small proportions come from coal and diesel power plants situated in the industrial areas. On the other hand, electricity demand in this region comes from the domestic, municipality, industry and agriculture use. The development of solar electricity was one of the main topics in the discussion either in the form of roof top or larger solar farms. With a considerable potential, renewable energy is expected to increase local electricity production and at the same time improving the ecosystem services in the region compare to other fossil-based electricity. In terms of access, this region has proper electricity infrastructure. However, the electricity tariffs are relatively unaffordable for low-income people. The development of solar energy is planned to be carried out in same locations as artificial ponds, namely in residential, industrial and agricultural areas within the region. So, the higher the development rate of those three sectors is expected to increase local solar energy production. It is noted here that the use of crop waste for household fuels was not mentioned during the GMB process. It was therefore assumed negligible, and was not captured in the developed CLD. The use of wood for fuel is implicitly captured in the energy sub-model (local energy production), with 6.77% of households using such fuel sources (BPS of Karawang, 2016b).

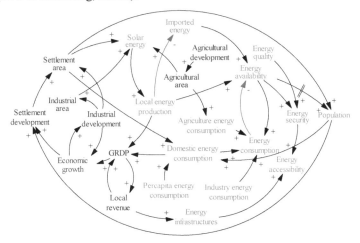

Figure 4.5. Energy security sub-model

c. Food security sub-models

The food security sub-model comprises availability, accessibility and quality of food produced in this region (Figure 4.6). Three types of food sources can increase food supply in this region, namely food-crops, livestock and fisheries products. On the other side, food demand generally comes from domestic and non-domestic consumption such as animal feed, seeds and raw materials for industry. The lack of food availability is met by bringing in supplies from outside the region and is classified as imported food. The main focus of the debate during the modelling session was related to agricultural land conversion. Since 1989, Karawang Regency has been established as one of the industrial development areas in Indonesia through Presidential Decree No. 53 of 1989 concerning industrial zones. Agricultural land in the region has shown a massive conversion either directly converted into industrial areas or indirectly into settlements, trade and services and infrastructures. The existence of regional regulations related to the land protection policy is expected to reduce the rate of land conversion in this area. In addition to availability, the issue of food accessibility and food quality are also an important to consider in achieving food security targets.

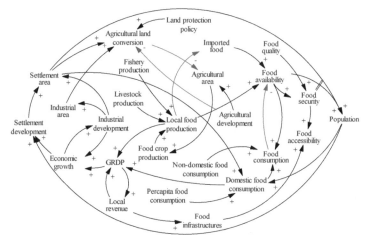

Figure 4.6. Food security sub-model

4.3.5 Karawang WEF security (K-WEFS) model

The above sub-models as endogenous drivers have been combined and integrated with other three exogenous factors in the second modelling session of the GMB workshop. The developed K-WEFS model is comprised of six sub-models i.e. water, energy, and food sectors as endogenous drivers, while population, economic, and ecosystem services were considered as exogenous drivers (

Figure *4.7*). Water-related variables in K-WEFS model are shown in blue text, while energy-related and food-related variables are represented by orange and green text respectively. In addition, purple variables indicates the exogenous variables. The relationship among variables in the system is indicated by black arrows with the positive "+" polarity and the red arrows with the negative "-" polarity.

The K-WEFS model is composed of 73 variables consisting of 20 food-related variables, 15 energy-related variables, 18 food-related variables and 20 external variables. There are three main foci of planned interventions to be carried out by the local government of Karawang to improve WEF security: (i) artificial pond development; (ii) solar energy development; (iii) regulatory enforcement on sustainable agricultural land to protect it from a massive conversion. In the K-WEFS model, starting from the "artificial pond" variable, 277 feedback loops are produced, while solar energy and agricultural land conversion formed 187 and 213 feedback loops respectively. In the next section, those three interventions will be further discussed practically in order to understand qualitatively how the interactions within the system are happening.

In general, the increasing number of residents and economic growth in the region will directly increase the demand for food, energy and water in terms of domestic and other uses. Conversely, the population and economic growth will also be influenced by the availability, access and quality of water, energy and food in this region. From an economic point of view, the long-term vision of Karawang Regency to achieve a prosperous region based on agricultural and industrial development may cause impacts to the local economic situation. In addition, the strategic location of Karawang which is close to the capital of Indonesia makes this region an important centre of economic activities besides industry such as residential, infrastructures, trade and services, hotels, and other supporting facilities. Such growth will potentially increase the interaction between water, energy and food sectors. Those two variables (population and economy) including ecosystem services are considered as external factors in the K-WEFS system.

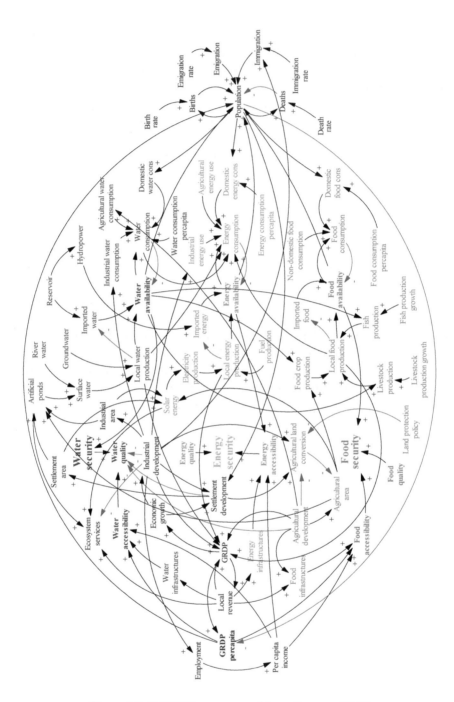

Figure 4.7. K-WEFS nexus causal loop diagram (CLD)

From the feedback loops indicated in the K-WEFS model, there is one reinforcing feedback loop (R) and one balancing feedback loop (B) that will be used as examples of model application to the Karawang Regency's development planning (Figure 4.8) to illustrate the advantage of systems thinking for planning purposes.

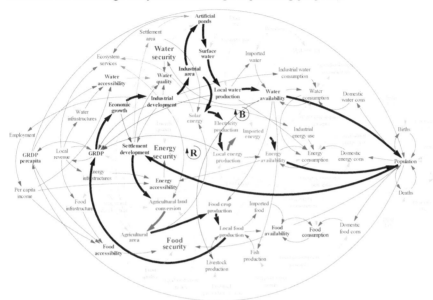

Figure 4.8. An example of K-WEFS model application in analysing qualitatively the planned interventions on water, energy, and food related sectors in the study area. Coloured variables, connected by thick arrows, highlight the feedback loops explicitly described in the text. Dark-shaded variables and arrows show all those variables affected taking a change in "Artificial ponds" as a starting point.

Balancing feedback loop (B)

Artificial ponds →(+) surface water →(+) local water production →(+) water availability →(+) population →(+) settlement development →(+) agricultural land conversion →(-) agricultural area →(+) food crop production →(+) local food production →(+) GRDP →(+) economic growth →(+) industrial development →(+) industrial area →(+) artificial ponds

---

One of the planned interventions in the Regency to increase local water production is the proposed development of artificial ponds. The increase of local water production is expected to reduce the water demand from the imported source (Jatiluhur reservoir). Based on data from Jasa Tirta Public Corporation II (PJT II) in 2011, the amount of water sourced from lakes or artificial ponds in this region was around 61.5 million m³/year from

33 scattered locations with a total area of 537 ha. The total volume is around 2.6% of the total potential water supply that flows in Karawang regency.

The construction of new artificial ponds in both urban and rural areas is expected to increase the capacity and volume of surface water availability, as well as contributing to mitigating the impacts of floods in the rainy season and offsetting shortages during the dry season (see loop B in Figure 4.8 and the box above). This will increase the local water production from surface sources and reduce the withdrawal of underground water that is currently critical in this region. The local government continues to make efforts by enforcing local planning and regulation to develop artificial ponds in residential, industrial and agricultural areas, especially rain fed agriculture, livestock and fishery areas. Increasing local water production will increase water availability in all seasons and all regions in Karawang Regency. However, on the other hand, the availability of water could also affect the population through mortality reduction and fertility enhancement due to basic needs of clean water fulfilment for the people in this region. A higher population would lead to a growth in settlement development, which increases residential areas. With the implementation of the artificial ponds development policy in particularly residential areas, the number of artificial ponds that are built may also significantly rise to serve the growing population, thus giving rise to this feedback process.

On the other hand, increasing settlement development leads to agricultural land conversion, causing an important impact to the area of agriculture, food-crop production and local food production. One of the factors influencing GRDP is local production, including food-related production (Purwanto *et al.*, 2018). However, local food production can be influenced significantly by land conversion. Karawang comprises irrigated paddy field (97,353 ha; 50.7%), non-irrigated agriculture area (10.5 ha; 5.5%), fish pond and aquaculture (18.79 ha; 9.8%) (Statistics Agency/BPS-Karawang 2015). The Local Regulation of Karawang No. 1 (2018) on protection of sustainable food crops farmland has set approximately 87,000 ha area in Karawang Regency as sustainable agricultural land, while another 10,000 ha are subject to be converted into other land uses until 2030. If the local government can consistently enforce the regulation, then the negative impact of land conversion to food-crop production in this region will be well-controlled. These balanced situations are important to be addressed by decision makers in Karawang Regency.

One measure to analyse economic growth is GRDP, where all outputs from economic activities in an area are calculated and considered. This covers all goods and services produced in an area. If the government of Karawang is able to maintain economic growth continuously, it may trigger investors to do their business in particular in residential and industrial sectors along with the increasing demand for goods and services. From 25 industrial estates that already have permits in Karawang Regency, only six of them are fully operated, while the rest have not carried out their operations, even the construction phase. Thus, if the regional government is to optimize artificial pond construction according to the local planning in industrial and housing estates through the regulation enforcement, the planned target to increase local water production is expected to be achieved. Such a growth in these sectors would lead to increased water demand, which could be serviced via the construction of yet more ponds, again forming a feedback relationship.

Reinforcing feedback loop (R)

Solar energy →(+) electricity production →(+) local energy production →(-) imported energy →(+) energy availability →(+) population →(+) settlement development →(+) agricultural land conversion →(-) agricultural area →(+) food crop production →(+) local food production →(+) GRDP →(+) economic growth →(+) industrial development →(+) industrial area →(+) solar energy

---

Another interesting interaction loop is reinforcing feedback loop (R; Figure 4.8, and box above), the main focus of which regards policy interventions in solar energy development. The availability and accessibility of energy is required to increase food production. The main issue that was raised in the topic of the energy sector is creating additional solar electricity via either roof-top installations or solar farms at any scale. The national energy target in 2025 comprises renewable energy of 23%, and around 6.5 GW of the target should come from solar-based electricity. The manufacturing industry consumes significant electricity, and the local drinking water company (PDAM) of Karawang uses electricity for water filtering and to purify raw water into drinking quality water. The increasing development of solar electricity in Karawang Regency will increase the local electricity production in a carbon-neutral way. In addition, its utilization as a complementary electricity resource can improve environmental conditions relative to fossil-based electricity generation. The construction of solar electricity in Karawang Regency can be carried out in industrial, residential, agricultural, livestock and fishery areas as a complementary source of electricity from the installed Electrical State Company (PLN). The increase in the use of electricity from solar energy is expected to reduce the electricity supply from PLN, so that it can be allocated for other purposes such as electric vehicles. In the long term, it will contribute to improving the environmental quality and reducing greenhouse gas emissions in the Regency.

The development of solar energy (roof top and large-scale solar farms) is expected to increase the overall availability of energy. In the feedback loop R (Figure 4.8 and the box above), the interconnection between solar energy and population is clearly indicated. The increase in population may promote the development of residential units in this region, and eventually could bring a positive impact to the development of solar energy in Karawang Regency by utilising the additional rooftop area for solar-electric generation.

In the same way, there is a causal feedback interaction that shows that development of new factories in this region may lead to increased development of solar energy for electricity, energy that can also be utilized by industry as an alternative energy sources to fulfil part of their electricity demand.

Two additional feedback loops (not explicitly shown in Figure 4.8) explain the impact of population and economic growth (i.e. exogenous drivers) on the WEF related sectors. As the number of residents in the region increases, it could lead to domestic water consumption increases as well as increases in food and energy consumption. The increase of water consumption, apart from increasing the pressure on these resources, will also raise the energy consumption in the region, because most of the water consumed comes from the sources that require electricity in the process of extraction, purification and distribution. The energy consumption increase could decrease the availability of energy resources in the region. These joint effects could feedback to impact future population in the region. The population in this region is not only influenced by birth rate and the declining of mortality, but also by migration from other regions in Indonesia. This is happening because of its attractiveness of economic development, especially the industrial sector development. In addition, the trade, services, hotels, culinary and construction sectors stimulate visitors and workers to come and settle in this region temporarily. Reductions in water, energy and food security due to large population increases (leading to supply stresses) could detrimentally impact in-migration to the region, which offers an important economic contribution.

In addition to the examples above, Figure 4.8 also highlights (in dark shading) all the nexus connections associated with a policy aimed solely at altering the area of artificial ponds. This highlighting indicates just how complex nexus interactions are in this Regency, and demonstrates the importance of whole-systems thinking and mapping. There are for example, indirect links from artificial pond area to energy consumption, links which may not even have been thought about or anticipated without such a nexus model.

The above discussion demonstrates the considerable added value of a qualitative causal loop mapping approach to better understand the complexity of the local water-energy-food security nexus, thus achieving the main objective of this study. Without recourse to computational modelling, stakeholders have greater appreciation of WEF nexus system complexity, and during the GMB process realised the potential impacts (positive and negative) of measures in their area of expertise across the whole nexus. Figure 4.8 shows that adjusting a single variable in the nexus could impact on more than 75% of other variables. Many of these connections are high-order and were unanticipated. Some feedbacks were shown to have consequences on their own sector, but were not considered initially due to the complexity. As a result, local planners are now better equipped to break silo thinking mentalities, and using such a nexus mapping can better understand the whole-system response to proposed objectives and measures. Through these examples, tracing the nexus using the K-WEFS model using real proposed plans and exogenous drivers in the Karawang Regency as a starting point, the added value of such a nexus mapping exercise is clearly demonstrated. Nexus-wide synergies and trade-offs are identified, many of which may not have previously been thought of. As a result, planners can act to modify potential plans and objectives to exploit potential synergies while at the same time minimising detrimental trade-offs, or where this is not possible, prioritising actions such that those with negative impacts are given less priority. Therefore, nexus-wide benefits can be achieved while minimising detrimental effects. Measures can be prioritised for synergy. Future studies can seek to 'convert' the causal mapping developed here into a quantitative computational model in order to get insight into the magnitude of potential impacts, however such an exercise would require a considerable volume of data.

Regarding previous studies, this work contrasts with many of those examples and adds scientific and practical value to them, thereby providing complementarity of approaches. For instance, the research conducted by Ma *et al.* (2019), Liang *et al.* (2019), Daher and Mohtar (2015), Bala *et al.* (2014) and other quantitative and qualitative nexus analyses will be more comprehensive if stakeholder engagement process is also included either in the beginning or in the end of the study. On the other hand, these findings further support the previous studies such as Inam *et al.*, (2015), Kotir *et al.*, (2017), Rich *et al.*, 2018), etc. that underlined the effectiveness of GMB to improve problem understanding and decompose the a complex system.

4.4 CONCLUSION

GMB and other participatory modelling approaches held in local institutions may have its own challenges compared with the same processes conducted in an academic environment. It is not always easy to handle dominant personalities, strong opinions and conflicting point of views from different participants. In this regard, the facilitator and modeller play an important role in ensuring the whole processes of modelling take place properly according to the agreed initial objectives. Open questionnaires help participants in grasping the topic and methods discussed in the workshop. In this case study, all participants agreed about the effectiveness of system dynamics GMB in analysing the policies and finding the best solutions not only in this topic but also in other development programmes, and see the value of CLD development for more coherent policy making, all without the need for quantitative model development. System dynamics GMB is important in improving better understanding of the problem, aligning perception, building strategy to improve performance of the WEF system, and developing more comprehensive qualitative-based model due to the involvement of multi-disciplinary experienced group of researcher, planner and policy-makers during its development.

The developed K-WEFS model provides an extensive and coherent overview of the complex water, energy and food systems in the Karawang Regency, Indonesia. All the feedback causal loops and variables can be used as a consideration not only by all related stakeholders in GMB workshop but also other stakeholders in any levels and scales to address all potential impacts that may occur. In addition, this model can also map clearly the positions, tasks and authorities that should be owned by each stakeholder. It will strengthen the efforts to carry out the institutional reforms and avoid overlapping authorities to achieve WEF security targets in the future.

This study has demonstrated the effectiveness of qualitative causal loop mapping of the WEF nexus in a GMB setting in decomposing a complex system by involving all related stakeholders. The added value of a qualitative approach, without recourse to computational modelling, in better understanding a complex system is clearly demonstrated. As more studies of this nature are developed, the opportunity arises for more general statements and conclusions for be formulated. Further study is recommended to be undertaken in investigating the interdependences of each variable and feedback loop in the K-WEFS model using a quantitative approach if so desired, although this will require considerable data. Quantifying this K-WEFS qualitative model is expected to bring more added value for the users in simulating the behaviour of the system over time, analysing every variable in the model more precisely, and getting an idea of the magnitude of potential changes.

5

QUANTITATIVE SIMULATION OF WEF SECURITY NEXUS

Abstract

The process of planning and evaluation for local development, especially in the critical sectors of water, energy and food (WEF) should be conducted using a holistic, integrated approach in an attempt to bring the improvement in water, energy and food security in a region. System dynamics models are one of the tools for simulation and assessment of the system-wide impacts caused by local interventions. This research develops a stock-flow diagram (SFD) of WEF security in a local context to be used in analysing the impacts of implementing three planned policy interventions in Karawang Regency, Indonesia. STELLA Professional software is employed to build the SFD and conduct simulation of the WEF security nexus, and is based on a previously developed qualitative causal loop model of the same system (the Karawang WEF security (K-WEFS) model). In the quantitative SFD, four scenarios are developed and assessed in this study; (i) population growth changes; (ii) agricultural land conversion rate changes; (iii) changes in the development of artificial ponds and solar energy; and (iv) per-capita resource consumption changes. The results show several interesting findings related to the WEF security nexus, available resources per person (APP) and self-sufficiency levels (SSL) of resources in business as usual conditions and under planned interventions. Potentially unanticipated detrimental indirect impacts of policy interventions are highlighted. This dynamic support tool could be applied in other local regions to improve the evaluation and planning process of water, energy and food sectors in a holistic manner.

Keywords: system dynamics, water-energy-food (WEF) security, nexus modelling, policy analysis, evaluation and planning, Karawang, Indonesia

This chapter is based on:

Purwanto A., Sušnik J., Suryadi F.X., de Fraiture C., (2020), *Quantitative simulation of the water-energy-food (WEF) security nexus in a local planning context in Indonesia*, Sustainable Production and Consumption Journal, Elsevier, DOI: https://doi.org/10.1016/j.spc.2020.08.009 (Published)

5.1 INTRODUCTION

Integrated management of water, energy, and food (WEF) is essential for human life and resource sustainability. Food production, food accessibility and food quality are three key elements in determining food security. Similarly, water security is described as the condition where people in a region are able '(1) to safeguard sustainable access, (2) to adequate quantities of acceptable quality water for sustaining livelihoods, human well - being, and socio-economic development, for ensuring protection against water-borne pollution and water-related disasters, and for preserving ecosystems in a climate of peace and political stability' (UN-Water, 2013). In terms of energy security, the International Energy Agency (IEA) defined it as 'an uninterrupted availability of energy sources at an affordable price'. All three resources must be managed in an integrated way, accounting for mutual dependencies. For example, De Fraiture and Wichelns (2010) underline that increasing water demand will escalate competition with the rising need of water for agriculture. Globally, energy demand will increase almost twofold, while the demand of water and food are foreseen to escalate by more than 50% in 2050 (IRENA, 2015). Due to the interconnectedness of the WEF nexus (de Fraiture *et al.* 2014; Endo *et al.* 2015; El Gafy *et al.*, 2016), a coherent and holistic modelling approach is best suited to explore nexus behaviour, especially in response to various policy objectives, and climate and socio-economic developments.

System Dynamics Modelling (SDM) is defined as 'the investigation of the information-feedback characteristics of (managed) systems and the use of models for the design of improved organizational form and guiding policy' (Forrester, 1961). SDM is used to analyse complex feedback-driven systems by imitating the system to the level of detail required (Sušnik *et al.*, 2012) and is used as a practical tool to assist policy makers in solving challenges in their organizations (Sterman, 2000). The SD modelling process begins with the construction of conceptual models and Causal Loop Diagrams (CLDs), which are then 'translated', quantified and simulated using Stock and Flow Diagrams (SFDs) (Sterman, 2000). CLDs and SFDs in system dynamics analysis are complementary. CLD gives qualitative understanding on the system structure and of the main connections between system components, and are very useful for understanding qualitative behaviour and the potential nexus-wide impacts of imposed (policy) changes (Purwanto *et al.*, 2019). SFDs are the process of quantification and simulation of the system using CLDs as a start point for model development. Stakeholder participation is sometimes used to ensure the level of complexity of the issues to be analysed.

SDM therefore offers a holistic approach to investigate WEF systems. SDM research related to WEF security has been carried out at many scales and from many perspectives. Prasad *et al.* (2012) develop a framework considering climate change impacts in rural areas in South Africa, developing tools and using locally-based scenarios. Zhang and Vesselinov (2016) built a model for predictive analysis to define trade-offs between sectors (water-

78

energy-food-environment) and assess cost-effectiveness in planning, strategies and policies. Water-related research has been conducted by several scholars (e.g. Sušnik *et al.* (2012, 2013); Sun *et al.* (2015), Kotir *et al.* (2016)). Generally, they apply causal loop diagrams (CLD) and stock-flow diagrams (SFD) in parallel to model feedback process, dynamic behaviour, and interactions among sectors including water resources, population, domestic, industry and agriculture. SDM is successful in helping stakeholders in mapping problems and managing water resources (Martinez *et al.*, 2018). Bala *et al.* (2014) modelled causality among variables in a rice system, and delays and non-linearity were analysed to assess complex and problematic conditions regarding food security in Malaysia. In the energy field, Feng *et al.* (2016) applied SDM to model water supply, power generation and environment relationships in Hehuang region, China, finding close connections between sectors' future trajectories. Pan *et al.* (2017) develop an SDM to assess the oil supply system in China in terms of over-capacity and energy security. Research by Daher and Mohtar (2015) on the "Water-Energy-Food Nexus Tool 2.0" describe a tool for analysing the WEF nexus, and evaluated and demonstrated the use of the tool for decision-making guidance in Qatar. Guo et al. (2001) simulated the complex interactions in Lake Erhai basin, China, supporting planning system for the basin. (Bakhshianlamouki *et al.*, 2020) develop and apply an SDM to explore the potential impact of proposed restoration measures in the Urmia Lake Basin, Iran, demonstrating that some measures may have detrimental impacts, and that careful selection and combination of other measures could prove more useful in achieving habitat restoration in the long term.

Despite the progress made, there is still a relative lack of research to explore the potential impacts of real policy directions and national objectives across all nexus sectors in a coherent way, and using such results to thus inform effective policy formulation. Hoff (2011) underlined the lack of harmonized 'nexus database' or analytical framework for monitoring or trade-off analyses as one of knowledge gaps in WEF security nexus. Additionally, there is no sole technique that able to be applied for every specific circumstance suitably (Endo *et al.*, 2015). To deal with different and specific situations in different regions, deconstruction of the nexus approach (Lele *et al.*, 2013) and specific context elaboration (El Gafy *et al.*, 2016) have to be considered in order to make more effective and contextualized solutions to assist decision makers in managing WEF resources. This results in a continuation of sectoral approaches to policy making, when it is clear that coordinated efforts to needed so that many goals can be simultaneously reached. This study starts to fill this knowledge gap by developing and applying a coherent, quantitative SDM for WEF security nexus in a local planning context to analyse the implications of planned policy interventions in water, energy and food based on Indonesian national ambitions. The model is applied in the Karawang Regency, Indonesia.

5.2 METHODS

5.2.1 Stock Flow Diagrams (SFDs)

SFDs comprise stocks, flows, auxiliary variables (also known as convertors) and definition of the system boundary (Binder *et al.*, 2004). Stocks accumulate material, and change due to material flowing into or out of the stock. The flows are functions that move material into and out of stocks (Sterman, 2000). Convertors act to influence flow rates. These objects are linked by connectors, which transfer information in the model, and form feedback loops. Due to complex structures, feedback, delay and non-linearity emerge (Sterman, 2000). Exponential growth (positive feedback), goal seeking (negative feedback), and oscillation (dynamic equilibrium) are the most basic behaviours modes observed. Other typical dynamics that also commonly occur are S-shaped growth, growth with overshoot, and overshoot and collapse (Figure 5.1). In this study, SFD representation of the Karawang WEF nexus was implemented in STELLA Professional® (www.iseesystems.com), a dedicated SDM modelling framework. Therefore, all modelling languages, notations and formulas are based on the STELLA system.

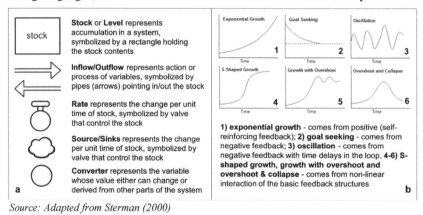

Source: Adapted from Sterman (2000)

Figure 5.1. a) Main elements of stock-flow diagram, and b) basic modes of dynamics behaviour

5.2.2 Study area

Indonesia has considerable resources to achieve water, energy and food security, including oil, coal, natural gas, high solar power potential, abundant water resources, and land resources to produce food. However, considerable attention has been paid to the level of security in Indonesia regarding water, energy and food, and how these resources are not being optimally managed, leading to scarcity situations. For example, the Asian Development Bank (ADB) ranked the National Water Security Index of Indonesia as 27 out of 48 Asian Countries (ADB, 2016b). An unfavourable ranking was obtained regarding energy security, where Indonesia ranked 85 of 125 countries in the Energy Trilemma Index (WEC 2016). Indonesia ranked 71 out of 113 countries for food security, behind other Southeast Asian countries like Singapore, Malaysia, Thailand and Vietnam (EIU 2016), while in the Global Hunger Index 2016, Indonesia was categorized as 'serious' (IFPRI, 2016).

The challenge of resource security is increasing, particularly in countries with decentralized governance systems such as Indonesia, where local government institutions have strong authority in managing natural resources, planning, and utilizing land use. Quincieu (2015) emphasizes the need for clearer roles, responsibilities, programmes and policies among district, provincial, and central government in Indonesia. Local governments have a significant role in determining the achievement of national targets in WEF security. Thus, the process of evaluation and development planning should be managed optimally at the regency (local) scale where decision making happens, but within the context of nationally-set objectives.

Karawang is located downstream of the Jatiluhur Reservoir, West Java Province, Indonesia (Figure 5.2). Geographically, Karawang is situated between 107°02'–107°40' East Longitude and 5°56'–6°34' South Latitude. Karawang Regency consists of 30 sub-districts, 297 villages and 12 special villages. Karawang Regency plays an important role as a major rice producer in Indonesia. Beside rice production, this region focuses on industrial development as stated in its long-term (twenty-yearly) planning for 2005-2025. The population of Karawang Regency in 2018 was 2,336,009 with a population density of 1,332 people/km^2 and a population growth between 2010-2018 of 1.17% (BPS of Karawang, 2019).

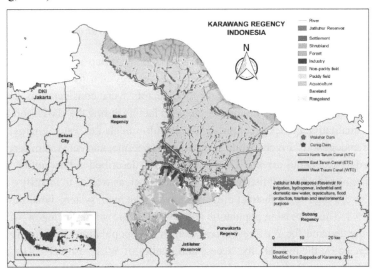

Figure 5.2. Karawang Regency map

81

5.2.3 Model development process

The WEF security nexus concept introduced by Hoff (2011) was used as the main reference in developing the system dynamics model. Available water resources as the central of the WEF system play an important role in supporting the security of other resources. By understanding the endogenous sectors of WEF and exogenous variables such as population growth, urbanization, climate change, economic development and improving all the enabling factors, the WEF security, sustainable growth and citizen resilience can be promoted. Figure 5.3 outlines the main stages in developing both the qualitative and quantitative Karawang WEF security (K-WEFS) nexus models.

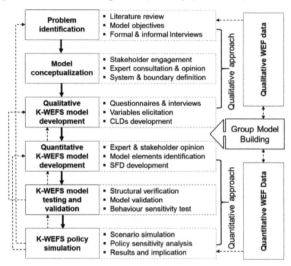

Figure 5.3. K-WEFS model development stages

The qualitative modelling process is characterized by group model building (GMB) involving local stakeholders related to the water, energy and food sectors to build an integrated CLD. The water, energy and food sectors were considered as endogenous sectors, and were integrated with three exogenous factors during the GMB workshop. The K-WEFS qualitative model is comprised of six sub-models i.e. water, energy, and food sectors as endogenous drivers, while population, economic, and ecosystem services were considered as exogenous drivers. The K-WEFS CLD described in Purwanto *et al.* (2019) was used as a guide for quantitative SDM development, as well as to identify data needs. The full here, a high-level abstract diagram showing the main system dynamics is shown in Figure 5.4. The increasing population and economic growth in the region will directly increase the demand for WEF resources. Conversely, population and economic growth are influenced by the availability, access and quality of WEF resources. The long-term vision of Karawang Regency to achieve a prosperous region based on agricultural and industrial development may bring impacts to both endogenous variables of WEF and exogenous variables such as population, economy and environment quality.

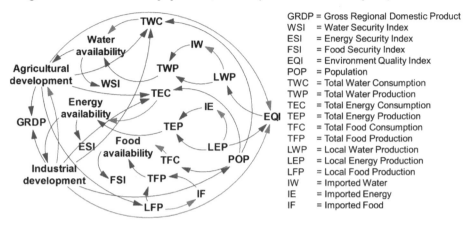

Figure 5.4. High-level dynamics mechanism of WEF security nexus in Karawang Regency, Indonesia. The blue arrows represent positive causalities (i.e. a change in variable X causes a change in variable Y in the same direction), while the red arrows indicate negative causalities (i.e. a change in variable X causes a change in variable Y in the opposite direction)

The quantitative SDM is represented by the development of the K-WEFS SFD in STELLA Professional. The SFD was developed to mimic as closely as possible the CLD such that examination of the implications of several planned policy interventions could be assessed. Data from 2010 was used as the base year in the analysis, and the model is run starting in 2010 and ending in 2030, with an annual time step. Table 5.1 shows the main initial data and parameters in the K-WEFS model, while all remaining data, including terms, definitions, and equations are presented in the Supplementary Material.

Table 5.1. Initial data of K-WEFS stock flow diagrams (base year 2010)

Variables	Initial value	Unit	Source
Population	2,127,791	people	BPS of Karawang (2011)
Total GRDP	99,641,319	million Rp	BPS of Karawang (2011)
Agriculture area	108,695	ha	Statistic of Agriculture Land 2014-2018, MoA (new release)
Housing area	24,121	ha	Rafiuddin et al. 2016
Industrial area	7,440	ha	Spatial Planning of Karawang 2011-2031
Aquaculture area	18,748	ha	Rafiuddin et al. 2016
Forest area	7,104	ha	Rafiuddin et al. 2016

Variables	Initial value	Unit	Source
Other area	25,919	ha	Rafiuddin et al. 2016; Spatial Planning of Karawang 2011-2031
Available land	191,864	ha	Rafiuddin et al. 2016

Two main resource indices are evaluated in this study: resource availability per person (APP); and resource self-sufficiency level (SSL) (Table 5.2). These indicators highlight the level of availability and the level of local resource production in the study area respectively. In addition, the Ministry of Environment of Indonesia established an environment quality index (EQI) to measure the performance of environmental protection and management (MoE-RI, 2018). EQI consists of three main components: a water quality index (WQI); air quality index (AQI/IKU); and land cover quality Index (LCQI). Two variables, WQI and AQI, are constants due to data limitations. Both water and air quality can be influenced by industrial development and other economic activities as captured in the K-WEFS CLD (Purwanto *et al*. 2019). The Water Security Index (WSI; *Table 5.2*) is adopted from the Water Security Framework developed by the Asian Development Bank (ADB, 2016b) and modified from the River Basin Water Security Index by PUSAIR, Ministry of Public Work (Hatmoko *et al*. 2017). It comprises five parameters (household WS, urban WS, economic WS, environmental WS and resilience to water-related disasters). The Energy Security Index (ESI) is modified from the RAND-WEF security index (Willis *et al*., 2016) consisting of energy availability and energy accessibility. The Food Security Index (FSI) is modified from the Food Security Agency, Ministry of Agriculture of Indonesia (BKP-RI, 2018) with three main parameters: food availability, food accessibility and food utilization. A weighting approach is used to determine the relative importance of the indicators. The weight for each indicator reflects the significance of the indicator in the security index. The score of each index is calculated within the range of 0-5, where 0 indicates poor, while 5 is excellent in terms of resource security. Unfortunately, due to data limitations at the local level regarding WSI, ESI, FSI and EQI, no calibration measures were done to these indices, and they are not included in the results. With better data, they can easily be incorporated however. Full details of each index are found in the Supplementary Information.

Table 5.2. Indices in K-WEFS stock-flow diagrams

No.	Index	Definition	Reference
1	Availability per person (APP)	Ratio between the resources supply (including imported resources) and total population	Own analysis
2	Self-sufficiency Level (SSL)	Ratio between local resource production and resource consumption or demand	Own analysis
3	Water Security Index (WSI)	Household WS, urban WS, economic WS, environmental WS	Adopted and modified from (ADB, 2016b)

No.	Index	Definition	Reference
		and resilience to water-related disaster. Score: 0-5	and (Hatmoko *et al.* 2017)
4	Energy Security Index (ESI)	Energy availability and energy accessibility. Score: 0-5	Modified from RAND-WEF security index (Willis *et al.*, 2016)
5	Food Security Index (FSI)	Food availability, food accessibility and food utilization. Score: 0-5	Adopted and modified from Food Security Agency, MoA (BKP-RI, 2018)
6	Environmental Quality Index (EQI)	Water quality index (WQI), air quality index (AQI), and land cover quality Index (LCQI). Score: 0-100	Adopted and modified Ministry of Environment (MoE-RI, 2018)

Verification of the model structure, model result validation and sensitivity analyses were performed to ensure the fitness of the model to observed data. Policy simulation is the final step. Using the verified model, four policy scenarios have been developed and applied to examine the implications of those interventions to the APP and SSL of water, energy and food.

5.2.4 Model validation and policy sensitivity simulation

a.   Model behaviour validation

To validate model results (X_m) against observed and statistical data (X_d), six measures as described in Table 5.3 were used, as suggested in Sterman (2000) as best practice for SD modelling.

According to Sterman (2000), several statistical methods can be used for assessing model fitness to the observed data; R^2, MAE, MAPE, MAE/Mean, RMSE, and Theil's inequality statistics. In addition, Kotir et al. (2016) applied three model behaviour tests namely M, R^2 and U_0. The explanation of each measure can be seen in Table 5.3. They suggested also to consider and analyse more on the trend and pattern rather than only predicted points resulted in that statistical analysis. In this chapter, in order to validate model results (X_m) against observed and statistical data (X_d), six measures as described in Table 5.3 were used.

Table 5.3. Summary of statistical measures to test the model behaviour

No.	Statistical Measures		Definition
1	Maximum relative errors (M) (%)	$M=\dfrac{\sum (X_m - X_d)}{\sum X_d}$	Maximum possible divergence between model (X_m) and data (X_d)

No.	Statistical Measures		Definition
2	Coefficient of determination (R^2) (dimensionless)	$R^2 = \left(\dfrac{COv(X_m - X_d)}{\sigma X_m - \sigma X_d}\right)^2$	Score 0 to 1 (closer to 1 indicating well-fitted) COv = covariance; σ = standard deviation
3	Discrepancy coefficients (U_0) (dimensionless)	$U_0 = \dfrac{\sqrt{\sum (X_m - X_d)^2}}{\sqrt{\sum X^2_m} + \sqrt{\sum X^2_d}}$	Score 0 to 1 (closer to 0 indicating perfect condition)
4	Mean absolute error (MAE) (unit)	$MAE = \dfrac{1}{n}\sum |X_m - X_d|$	The average of the absolute errors and weights all errors linearly
5	Mean absolute percent error (MAPE) (dimensionless)	$MAPE = \dfrac{1}{n}\sum \left|\dfrac{X_m - X_d}{X_d}\right|$	Provide dimensionless metrics for the error and indicate the prediction accuracy
6	Root mean square error (RMSE) (unit)	$RMSE = \sqrt{\dfrac{1}{n}\sum (X_m - X_d)^2}$	RMSE weights large errors much more heavily than small ones

Source: adapted from Sterman (2000); Kotir et al. (2016)

b. Sensitivity analysis

In this study, policy sensitivity on birth rate are applied to the model using Monte Carlo simulation and incremental distribution provided by Stella Professional®. Dynamic confidence intervals can be generated using monte-carlo simulations for the trajectories of the variables in the models (Sterman, 2000).

The number of total simulation samples is 1,000 runs for three sensitivity parameters i.e. birth rate, industrial growth, and agriculture shrink rate and with all starting and ending value of 0 and 1 respectively. By specifying incremental distribution, it will automatically create values that incremented evenly from the start to the end values and then distributed over the number of simulation samples with the confidence interval of 50%, 75%, 95% and 100% including the mean value of six parameters to be tested namely APP and SSL of water, energy and food.

c. Policy analysis

There are four policy scenarios that have been proposed by policy makers and other stakeholders in the study area which will be analysed in this study. These four comprise of: (1) population growth assumptions; (2) differing levels of agricultural land conversion; (3) changes to artificial pond development and solar energy development; and (4) changes to resource consumption patterns. Table 5.4 shows the main features of the scenarios applied to the model. Each category is explained below.

Table 5.4. Policy and scenario analysis

No	Scenario	Baseline (2010)	Change			
			↑20%	↑40%	↑60%	↑80%
1	(SCE#1) In migration rate increase	SCE#1.0 0.017	SCE#1.1 0.0204	SCE#1.2 0.0238	SCE#1.3 0.0272	SCE#1.4 0.0306
2	(SCE#2) Agriculture land conversion rate	SCE#2.0 0.0071	SCE#2.1 0.0085	SCE#2.2 0.0099	SCE#2.3 0.0114	SCE#2.4 0.0128
			↑200%	↑300%	↑400%	↑500%
3	(SCE#3) Artificial pond development growth	SCE#3.0 0.001	SCE#3.1 0.003	SCE#3.2 0.005	SCE#3.3 0.007	SCE#3.4 0.009
	Solar energy development growth	0.001	0.003	0.005	0.007	0.009
		SCE#4.0	SCE#4.1	SCE#4.2	SCE#4.3	SCE#4.4
			↑20%	↑30%	↑40%	↑50%
4	(SCE#4) Percapita water consumption (m³/cap/y)	55	66	72	77	83
	Percapita electricity consumption (kWh/cap/y)	634	761	824	888	951
	Percapita staple food consumption (ton/cap/y)	0.118	0.142	0.153	0.165	0.177

Scenario #1: Population increase

The first scenario is related to population growth. As one of the satellite cities of Jakarta, economic growth of Karawang is relatively high. Aside from being supported by industrial development, its strategic location makes this region a main destination for business, trade and services expansion, including infrastructure development. This situation attracts people to the region. In migration rate (i.e. the number of residents coming into Karawang) is predicted to increase over the increase of economic development, in particularly industrial, trade and services development in this region. By 2010, the rate of in migration was 0.017. Increases in this rate by 20%, 40%, 60% and 80% from the base year were evaluated to examine the impacts to the APP and SSL of water, energy, and food. This scenario explores the implications of the growth of urbanization.

Scenario #2: Agricultural land conversion

In scenario 2, agriculture land conversion was considered. This issue is becoming a major concern not only for the local government of Karawang but also for provincial and national government. The average agriculture conversion rate in Karawang is around 0.71% to 1.43% per year (Spatial Planning of Karawang 2011-2031; Rafiuddin *et al.* 2016). Most of the agricultural land was converted into residential and industrial areas (Nadia Putri Utami and Ahamed, 2018; Widiatmaka *et al.*, 2013). The changes in this scenario consider increases of 20% to 80% from the base year.

The Local Regulation of Karawang No. 1 Year 2018 on the protection of sustainable food crops farmland has set a target of approximately 89,411 ha area in Karawang Regency as sustainable agricultural land, while the remaining areas are subject to be converted into other land uses until 2030. If the local government can consistently enforce the regulation, the negative impact of land conversion to food-crop production in this region will be well-controlled. These balanced situations are important to be addressed by the decision makers in this region for overall resource security.

Scenario #3: Artificial pond and solar energy development

Scenario 3 is designed to examine the implications of planned interventions in the construction of artificial ponds and solar electricity on the APP and SSL of water, energy and food. As these interventions are not major government initiatives, the growth is designed to increase significantly relative to a very low starting point (i.e. from 200% to 800% of the base year value of 0.1%). Artificial ponds development in both urban and rural areas is expected to increase the capacity and volume of surface water availability, as well as contributing to mitigating the impacts of floods in the rainy season and offsetting shortages during the dry season. This will increase the local water production from surface sources and reduce the withdrawal of underground water that is currently critical in this region as stated in the long-term planning (RPJPD) of Karawang 2005-2025. The local government continues to make efforts by enforcing local planning and regulation to develop artificial ponds in residential, industrial and agricultural areas, especially for rain fed agriculture, livestock and fishery areas. Increasing local water production will increase water availability in all seasons and all regions in Karawang Regency. From calculation, referring to the letter of Ministry of Public Work No. 7/2018, the potential of water from an artificial pond that can be utilized per ha area is about 3,000 m^3/ha by assuming 1,500 m^2 maximum area and 2 meter maximum depth.

Many provinces in Indonesia have high potential for solar power output, including West Java Province. This high potential can be optimized for meeting local energy needs, especially with the use of solar panels on rooftops in various buildings (IESR, 2019). The development of roof-top solar electricity is expected to increase the availability of local energy source, and may offset fossil-fuel dependant sources. The industrial development growth may promote the development of residential units in this region, and eventually could bring a positive impact to the development of solar electricity in Karawang Regency by utilising roof-top area for electricity generation. In the same way, electricity from solar can also be utilized by industries as an energy alternative sources to fulfil part of their electricity demand.

Scenario #4: Resources consumption

This scenario attempts to depict the impact of resource consumption relative to the APP and SSL of water, energy and food in the Regency due to the possibility of per capita resources consumption increase of 20% to 50% from the base year.

The relationship between income and consumption is used to describe economic trends in the household sector. When income increases, disposable income rises and thus people consume more goods and services, including water, energy and food consumption (Muhammad et al. 2017; Chang 2014; Gamage and Jayasena 2018). This is mostly caused by the change of people's life style and habits. Based on statistical data, per capita income of people in Karawang Regency is increasing over time (BPS of Karawang, 2018).

5.3 RESULTS

5.3.1 K-WEFS nexus model structure

The K-WEFS SFD consists of three endogenous sectors: water (blue), energy (orange) and food (dark green) (Figure 5.5). This quantitative K-WEFS SFD is based on the qualitative K-WEFS CLD that developed and fully described by Purwanto et al. (2019) and briefly described above (Figure 5.4). Exogenous factors driving changes in the WEF sectors comprise population, land, economic and ecosystem services (dark purple; Figure 5.5). The water sector consists of 48 converters and 2 stocks, while the energy and food sectors comprises 30 converters and 2 stocks, and 43 converters and 2 stocks respectively. The model contains 14 converters and one stock in the population sector, 11 variables and seven stocks in the land sector, seven variables and one stock in the economic sector, and nine converters in the ecosystem services sector. Pink variables labelled WI1 to WI13 (Water Indicator 1 to 13; Figure 5.5) in the water sector represent normalized values and weighting scores regarding water security (e.g. WI12 is a normalised index for water quality in the Regency). Similarly, EI1 to EI3 and FI1 to FI9 indicate Energy Indicators and Food Indicators regarding energy and food security. Each of these indicators is described in the Supplementary Information. The energy sector is limited to electricity production and consumption due to data limitations and because the authority of fuel management is under the control of the national government rather than local government. Not all of the variables in the K-WEFS CLD are covered in the SFD due to data limitations (e.g. infrastructure, fuel production, local revenue, and employment). However, there are variables in the SDM not in the CLD. The number of variables in the SFD is greater than compared with the K-WEFS CLD.

Supplementary data associated with this K-WEFS SFDs can be found in Appendix and downloadable through http://doi:10.1016/j.spc.2020.08.009 (Purwanto *et al.*, 2020b).

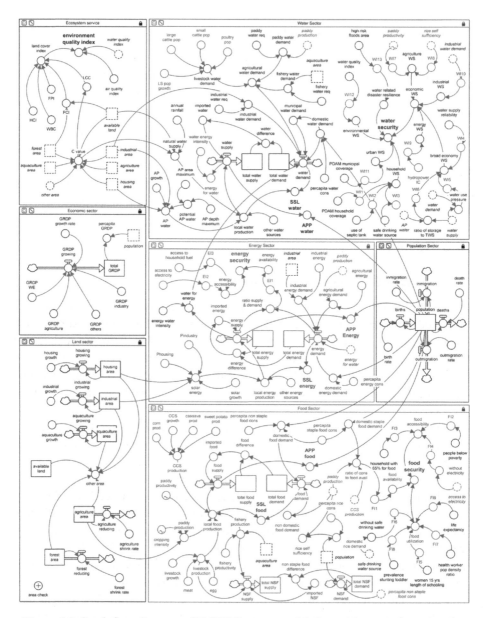

Figure 5.5. Stock-flow diagram of K-WEFS nexus model. Square boxes indicate stocks, thick arrows with 'clouds' indicate flows, and circles indicate connectors (auxiliary variables). Thin connecting arrows transmit information between model elements. This SFD is based on the CLD developed and fully described in Purwanto et al. (2019)

5.3.2 Model behaviour test results

The results of the baseline simulation are presented in Table 5.5 and *Figure 5.6* comparing simulations with observed data for the period 2010-2019.

Table 5.5. Model testing of selected variables (2010-2019)

Selected variable	Model testing					
	M (%)	R^2 (dmnl)	U_0 (dmnl)	MAE (unit)	MAPE (dmnl)	RMSE (unit)
Population (people)	2.56%	0.974	0.018	61,947	5.362	81,110
Agriculture land (ha)	1.35%	0.882	0.008	1,406	2.716	1,603
Paddy production (ton)	3.38%	0.419	0.021	49,228	6.858	62,654
Energy Supply (kWh)	2.74%	0.771	0.126	1,374,163,806	59.510	1,455,446,349

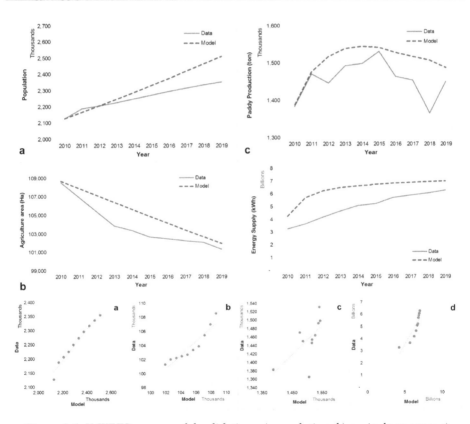

Figure 5.6. K-WEFS nexus model validation: a) population; b) agriculture area; c) paddy production; and d) energy supply

The model shows good agreement with observation, both in terms of absolute numbers and historically observed trends. This result can be considered as one of the validation methods demonstrating that the model is able to adequately capture the dynamics of the historical evolution of the WEF system in Karawang Regency, lending support for further validation test results and scenario analysis.

5.3.3 Quantitative K-WEFS model analyses under BAU conditions

Figure 5.7 shows the results using BAU assumptions, with the model running for 20 years from 2010 to 2030. The population is expected to increase to 3,077,850 people in 2030. Population increases alter water, energy and food availability per person (APP) until 2030 (Figure 5.7a). The APP water drops to 830 m^3/cap/y compared with 1,115 m^3/cap/y in 2010. A similar trend is expected in the food sector, where APP food decreases by 46% from 0.413 tons cap^{-1} yr^{-1} in 2010. Slightly different variations are shown in the energy sector where there is an increase in APP energy from 2010-2014, but after this a decreasing trend is observed from 2,901 kWh/cap/y in 2014 to 2,546 kWh/cap/y by 2030.

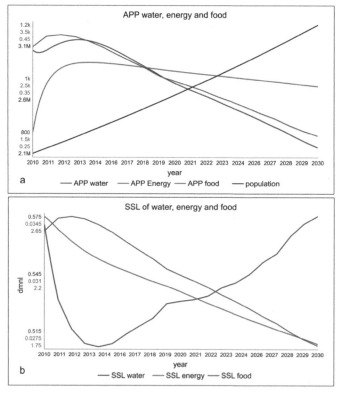

Figure 5.7. Model results of selected variables at base run: a) availability per person (APP); b) self-sufficiency level (SSL) of water, energy and food

Figure 5.7b shows the trends of water, energy and food SSL. The energy and food sectors have decreasing trends (SSL Energy from 0.034 to 0.028 and SSL Food from 2.553 to 1.774). In contrast, SSL water shows an initial decrease followed by an increasing trend from 0.516 in 2014 to 0.574 in 2030, most likely influenced by a declining irrigation water demand due to agricultural land conversion. The fluctuation of SSL values can be caused by the increase of local resources production and/or the decrease of resource consumption.

The K-WEFS model is able to capture variables related to water, energy and food security, and can calculate the relationship between variables to highlight the impact of planned interventions on the APP and SSL of water, energy and food in Karawang Regency. Major trends in water, energy, food, and ecosystem services under BAU conditions are shown in Figure 5.8.

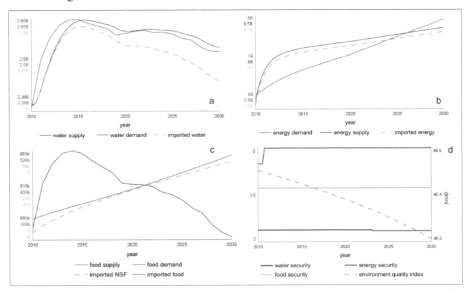

Figure 5.8. (a) water supply, water demand and imported water trends in m³/year, (b) energy supply, energy demand, and imported energy trends in kWh/year, (c) food supply, food demand and imported non-staple food in tons/year, (d) environment quality index, water security index, energy security index, and food security index (dimensionless)

a.   Water-related issues

The five main parameters of water security (WS) according to ADB (2016) and Hatmoko et al. (2017) include household, urban, economic, and environmental WS, along with resilience to water-related disasters. Here, the main issues of concern are related to water supply and demand. Water demand is dominated by industrial and domestic users. Groundwater extraction is permitted by regulation with several restrictions. However, other efforts are needed to meet local demand by utilizing surface water to reduce groundwater overexploitation.

The results of water availability (Figure 5.7a) can be related to the Water Stress Index or Falkenmark Index which consist of two main parameters: water shortage and water stress (Falkenmark *et al.*, 1989; Kummu *et al.*, 2016). The APP water is comparable to water shortage, while the ratio between water consumption and availability is equivalent to water stress. Under BAU conditions, in 2010 this region was in 'high water stress and medium water shortage' and tends to 'over-exploit' resources with an APP water of 1,115 m^3/cap/y. The situation worsens by 2030 without any intervention (APP water = 830 m^3/cap/y), suggesting decreasing water availability.

Figure 5.8*a* shows a decreasing trend of water demand related to agricultural land conversion. The stakeholders in irrigation water supply can make adjustments in allocating water for agriculture so that water for other users (industries, domestic users) can be fulfilled. At the same time, water supply is shown to decrease at a similar rate, compromising demand fulfilment. The water security score is in the range of 2.35-2.38 out of 5, indicating a lack of water security. This is despite predicted reductions in agricultural water demand. There are several other causes for the low WS score. First is the low piped water coverage to households which in 2010 was 11.43% (score 1) but increased to 23% (score 2) in 2019. The second factor is the low environmental WS related to water quality with a score of 1.

Another important aspect is a possible increase in per capita water consumption as living standards improve along with household connections. The quality of surface water is another major concern for policy makers. Water quality factors are important in influencing water security and environment quality (hence the low river water quality score of 1). Another interesting issue is the influence of energy and food sector variables in the water sector, including hydropower capacity, water use intensity, energy for water, paddy production, and rice self-sufficiency. Such connections demonstrate the interconnectedness with other nexus sectors.

b. Energy-related issues

The energy security components are less complex than water and food security, considering only availability and accessibility as local level data on energy quality was not available. The biggest electricity users in Karawang are industrial and business activities (81%), compared to households (18%), government buildings (0.5%) and social uses (0.6%) (BPS-Karawang 2016). Almost all supply in 2019 (±6.29 billion kWh) came from the electricity state company (PLN), while the rest was generated from local sources such as diesel, steam, coal and small solar power plants.

Figure 5.8b shows that under BAU conditions, energy demand will increase to 7.62 billion kWh by 2030, in line with population and economic growth, implying PLN must plan to provide additional electricity to meet most of the demand increase. Increasing electricity self-sufficiency through the development of renewable energy options should become a major issue for policy makers. In addition to reducing dependence on external sources, greenhouse gas emissions resulting from fossil-based energy sources can be reduced if renewable sources are increasingly used. The energy security score is 5 of 5 because the index only considers availability and accessibility without looking at level of self-sufficiency and local energy production. The linkages among endogenous variables are also seen in the energy sector, where water and food variables such as water for energy, energy water intensity and paddy production affect demand in the energy sector.

c.   Food-related issues

The index to determine food security refers to the Food Security Agency, Ministry of Agriculture of the Indonesia (BKP-RI, 2018) with adjustments to match available data. Food security is influenced by three main components: food availability, accessibility and utilization. The score for food security is 3.73 of 5. Food production is more than three times domestic demand, yet does not guarantee a maximum food security score because of other influencing factors, such as toddler stunting prevalence and the level of women's education that both need improvement. In the food sector, non-staple foods (NSF) such as livestock and fisheries products were included in the KWEFS model in addition to staple food.

 According to Rafiuddin *et al*. (2016), c. 960,000 tons rice were produced in 2010 in Karawang, while the domestic rice consumption was c. 450,000 tons, in line with the rice development roadmap of Karawang Regency 2007 which states that rice production should be in surplus and contributes to national rice sufficiency (Widiatmaka *et al.*, 2013). Food supply will gradually decrease due to land conversion from agricultural land (Figure 5.8c), a point for policy makers to take note of as it has national-level food sufficiency implications. Crop failure due to floods and droughts will reduce rice production, though the extent of such impacts was not evaluated in this study. Figure 5.8c indicates that there is no requirement to import rice and other staple food (corn, cassava, sweet potato) for domestic demand under BAU conditions. Nevertheless, the trend of imported non-staple food (NSF) such as fish and livestock products increases significantly until 2023 after which it levels off. Increasing the local production of non-staple food products is expected to reduce supply dependence from outside the region. Considering the local regulation of Karawang (Perda) no. 1 (2018), if the conversion rate of agricultural land to other uses is maintained at 0.71% per year, by 2030, the agriculture land will be around 94,294 ha, meeting the minimum target of 89,411 ha. However, by assuming the highest conversion rate in scenario#2 (1.28%), the remaining agricultural land is only around 84,111 ha (less than the target). Therefore, the rate of agricultural land conversion must be kept strictly in check with regulations and measures.

d.   Environment-related issues

Water security, energy security, food security and the environment quality index (EQI) are presented in Figure 5.8d. Water and air quality problems become a major problem due to the development of industry, transportation and settlements. The water quality index value is 20 out of 100 (single-year value). The low value comes from river water monitoring results of seven parameters (TSS, DO, BOD, COD, total phosphate, faecal coli and total coliform), with these being compared with national standards on river water quality and calculated into a pollution index (PI) (MoE-RI 2018; see Supplementary Information for details on the PI). The air quality index of 56/100 refers to the score of this index in the Annual Performance Report (LAKIP) in 2017 for Karawang Regency. The EQI gradually decreases from 40.4 to 40.2 of 100 in the period of 2010-2030. The EQI is almost constant with only a negligible change in values over the simulation period because it is only influenced by the land cover quality index. However, there are two additional parameters that influence EQI (water quality index and air quality index). An example can be seen in the change of land cover index due to land conversion. Although single figures are available for water and air quality, analysis of trends in water quality and air quality were not considered due to only a single data point being available. Development and other economic activities may cause negative impact to ecosystem services if it is not well managed. Ideally, there would be a quantitative feedback here as is suggested in the K-WEFS CLD, but a lack a data prohibited such quantification. The ecosystem service condition will influence and be influenced by the WEF system.

5.3.4 Sensitivity and policy scenario analysis

a. Sensitivity analysis

Based on simulations (SENSA-1a to 1f and SENSA-2a to 2f; Figure 5.9), changes in birth rate significantly alter APP water, energy, and food (Figure 5.9a-c). APP water (Figure 5.9a) and APP energy (Figure 5.9b) are very sensitive to changing birth rate. In the national mid-term planning (RPJMN), the target fertility rate (TFR) is expected to reach 0.0028, a decrease of 90% from baseline. Under an increase of 20% and 40% in birth rate however, APP water will decrease 5%-10%, APP energy decreases 4%-8%, and APP food decreases 6%-11% (Figure 5.9a-c). A decrease of 60% and 90% in birth rate from baseline, in line with RPJMN expectations, leads to APP water increases of 20%-29%, APP energy increases of 17%-24%, and APP food increases of 22%-32% (Figure 5.9a-c).

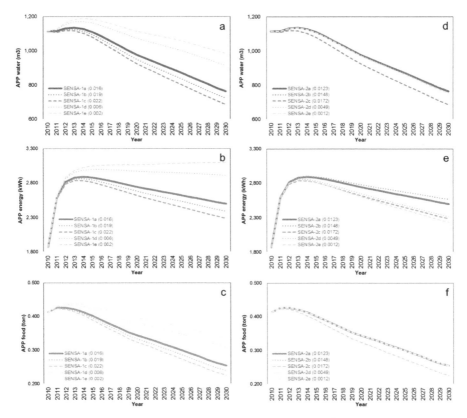

Figure 5.9. Sensitivity analysis of APP water, energy, and food parameter to the positive and negative changes in the birth rate (a-c) and industrial growth: (d-f). The bold lines in graph a-f represent the baseline results.

Figure 5.9 d-f show expected changes in APP water, energy and food due to changes in industrial growth rates. APP water decreases by 0.4%-10%, APP food decreases by up to 11% and APP energy increases by 3% due to increases of 20% to 40% in industrial growth rates. This is due to the growth in industrialisation is expected to increase solar electricity production and therefore raise APP energy slightly. Decreases of 60% and 90% in industrial growth only cause a decrease in APP water of 1%-2%, and a decrease in APP energy of 7%-11%. APP food does not experience changes with decline in industrial growth.

b. Scenario #1

Results of Scenario 1 (Table 5.4) show that changes in in-migration rate have an impact on APP and SSL of water, energy and food (Figure 5.10a-f). Significant reduction (18% to 24%) occurs in APP water, energy and food, as well as on SSL food to 2030 (Figure

97

5.10a, b, c and f). Reductions in SSL water and energy are relatively small, only 3% to 7% compared with the baseline (Figure 5.10d and e) demonstrating the unequal impact of the effects of in-migration. The difference between sectors is because food demand mostly comes from domestic consumption which is heavily affected by total population. In contrast, water and energy demand are mostly from agricultural and industrial activities in the region, and are less impacted by the changes in population (cf. Figure 5.5).

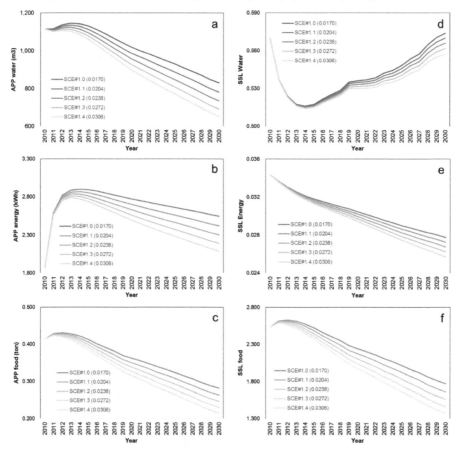

Figure 5.10. The implication of scenario #1 (in-migration rate increase) on the APP and SSL of water, energy and food in the period of 2010-2030

c. Scenario #2

Unlike in Scenario 1, APP and SSL of energy are not significantly affected by agricultural land conversion (decreasing around 2% by 2030; Figure 5.11) because energy demand within the agricultural sector is relatively less important (Figure 5.11b and e). The SSL of water and energy increase relative to the baseline (Figure 5.11d and e) by 2% to 10%

respectively from 2010 values. This result, especially SSL water, is explained by the fact that almost 70% of water supply to this region goes to irrigated agriculture (Figure 5.5). If agriculture land conversion continues, water demand to agricultural areas will drop, and the water supply authority can allocate this 'extra' water for other purposes.

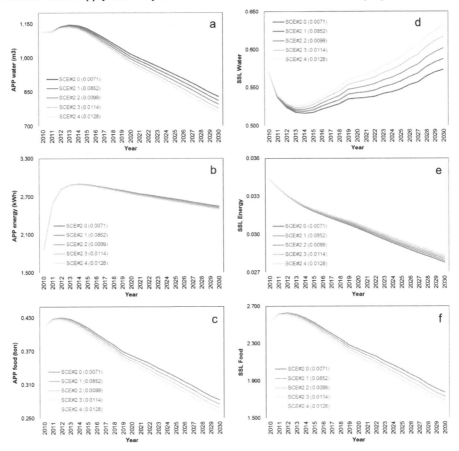

Figure 5.11. The implication of scenario #2 (agricultural land conversion) on the APP and SSL of water, energy and food in the period of 2010-2030

d. Scenario #3

The development of artificial ponds and solar electricity will significantly impact on SSL water (4%) and SSL energy (279%), while having a small influence on APP energy (0.3%), even with a significantly increased growth rate (Figure 5.12b, d, and e). The significant effect on SSL energy is because almost all electricity supply comes from the state electricity company, and only about 3% is generated from local sources. Meanwhile, the proportion of local water supply is relatively high (c. 50% of the total). One explanation is due to inadequate support from policy makers in terms of regulation and financial measures to link these two interventions with food production. If industrial and residential areas are able to develop more artificial ponds with government support, the potential of local water resources could be around 67 million m³/year (using BAU growth of 0.1%), and could contribute to additional food production, according to the system dynamics (Purwanto *et al.*, 2019), Figure 5.4, and Figure 5.5.

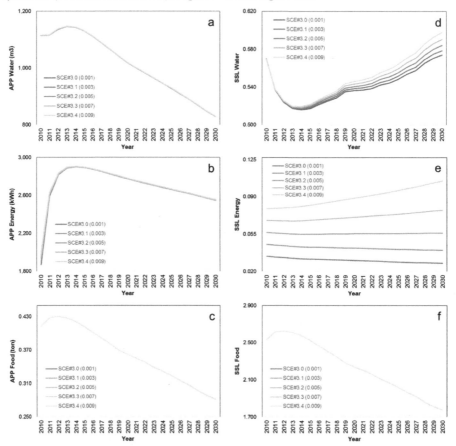

Figure 5.12. The implication of scenario #3 (artificial pond and solar electricity development) on the APP and SSL of water, energy and food in the period of 2010-2030

e. Scenario #4

The effect of increasing per-capita consumption of water, energy and food on APP and SSL is considerable (Figure 5.13). Almost all SSL parameters decrease with increases of between 20% and 50% in per capita consumption. Decreases in in the SSL water, energy and food of 5%, 11% and 27% respectively are modelled. Conversely, an increasing trend is shown in APP water and energy with increases of 5% and 13% respectively. APP food is influenced by food supply and population (Figure 5.5). However, due to the high staple rice production in this region, the increase in per capita consumption does not affect this parameter. This will be elaborated in the discussion.

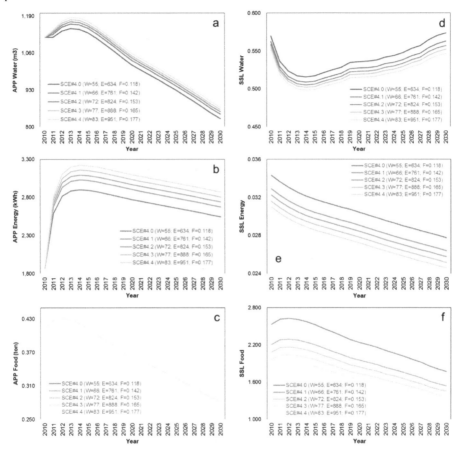

Figure 5.13. The implication of scenario #4 (water, energy and food per capita consumption) on APP and SSL of water, energy and food in the period of 2010-2030

5.4 DISCUSSION

The model accommodates variables related to water, energy and food and exogenous variables including population, economy and ecosystem services (which are affected by economic growth but which do not impact upon economic growth in this model), including their links and interactions. Better knowledge of such interactions and their impact on system response is useful for evaluation, planning, and decision making related to resource management and development. This model shows how proposed measures impact on the WEF nexus, highlighting trade-offs and synergies among the three main sectors in the study area. Table 5.6 summarizes the policies and their main implications regarding water, energy and food including practical actions to consider to reach targets and achievements set by the local government. It also shows two main concerns, first is "implication" (on indicator results) and second is "possible practical action to be taken". All results in "implication" are based on model simulation, while measures suggested in "possible practical action to be taken" come from both model simulation and from other considerations such as local planning ambitions, national programmes and local experts and modeller's opinion.

Table 5.6. Implication and practical recommendation on WEF-related policy and planning

Policy and planning	Implication	Possible practical action to be taken
Water Artificial ponds (AP) development growth (↑)	APP W (→)* APP E (→)* APP F (→)* SSL W (↑)* SSL E (→)* SSL F (→)*	1. Include in the mid and long-term planning to ensure the target's achievement** 2. Increase the growth of AP development* 3. Law enforcement and supervision on industrial, housing and other public facility developments** 4. Involve other stakeholders (not only water-related agency)** 5. Extend the purpose of AP to flood mitigation, leisure or tourism, fishery and other people activities**
Energy Solar rooftop electricity (SRE) development growth (↑)	APP W (→)* APP E (↑)* APP F (→)* SSL W (→)* SSL E (↑)* SSL F (→)*	1. Provide clear rules on SRE development & investment** 2. Educate people on the potential and importance of SRE** 3. Prepare a subsidy scheme for SRE* 4. Collaborate with PLN to increase the growth of SRE* 5. Involve other stakeholders (not only government institutions)**
Food Percapita WEF resource consumption (↑)	APP W (↑)* APP E (↑)* APP F (→)* SSL W (↓)* SSL E (↓)* SSL F (↓)*	1. Educate people on water, energy and food resources consumption and efficiency* 2. Propose food consumption diversification** 3. Prepare a subsidy scheme for substitute foods** 4. Educate people to consume a balanced and healthy diet*
Land & Environment Land conversion (↑)	APP W (↓)* APP E (↓)* APP F (↓)* SSL W (↑)* SSL E (↑)* SSL F (↓)*	1. Agriculture land conversion rate, until 2030, should be kept less than 1% per year to achieve the target in LP2B* 2. Law enforcement and supervision in land use permit and land conversion based on regional planning** 3. Increase agro-industrial development to stimulate a balance growth of industrial and agricultural sector** 4. Improve coordination with PJT II regarding irrigation water supply based on factual situation of agriculture area**

Policy and planning	Implication	Possible practical action to be taken
Population In migration rate (↑)	APP W (↓)* APP E (↓)* APP F (↓)* SSL W (↓)* SSL E (↓)* SSL F (↓)*	1. Increase the provision of WEF resource to meet demand* 2. Distribute the centre of economic growth by making collaboration with other regions to anticipate excessive urbanization** 3. Introduce vertical settlements to avoid land conversion from agriculture to the housing and trade sectors**
Economy Industrial growth (↑)	APP W (↓)* APP E (↑↓)* APP F (↓)* SSL W (↓)* SSL E (↓)* SSL F (→)*	1. Increase agro-industrial development to improve the value added of agricultural products** 2. Promote water and energy efficiency in industrial sector* 3. Increase water and energy production in industrial estate through AP and SRE policies* 4. Coordination with national and other local governments**

Notes: ↑ (increasing trend), ↓ (decreasing trend), → (less or no influence), PLN (state electricity company), PJT II (state water authority), LP2B (sustainable food crop agriculture land policy), AP (artificial pond), SRE (solar rooftop electricity), * (model- based), ** (other evidence-based)

Although artificial ponds (AP) have no significant effect on APP water, this policy does improve SSL water. Based on the model, the main water source for domestic purposes comes from the reservoir. Additionally, population growth in this region influences the APP water result since APP water is the ratio between water supply and population. However, according to the local government's experiences and local expert opinions during the process of qualitative group model building (Purwanto *et al.*, 2019), AP development may contributes in reducing flood events, improving the fishery sector, and increasing local income from tourism activities. Increasing the coverage of piped water by the local water supply company, law enforcement, and fines for violation of water quality regulations may also help contribute to improving water security in the region. On the other hand, there is no impact from the population to SSL water, thus the increase of local water production due to AP development significantly increases the ratio,

Solar rooftop electricity (SRE) policies have a considerable impact in increasing SSL energy (279%) and a slight change in APP energy (3%). The population in the region means that APP energy was not significantly increased. To improve on this, several practical steps have been set out in Table 5.6, such as providing clear regulations regarding SRE development and investment in various sectors. Collaboration with stakeholders regarding subsidy schemes, increasing the use of SRE and awareness raising are other actions that can be considered to maximise SRE uptake and therefore impact of this policy.

The increase of per-capita resource consumption causes a decrease of SSL values on one side, but an increase of APP on the other side. This is because SSL is defined as the ratio between local resource production and demand that is strongly influenced by per-capita consumption (Figure 5.5). APP however is defined as the ratio between resource supply and population, and as such, per-capita demand does not influence APP. APP food, water and energy increases because most supply comes from outside the region and is therefore under the control of the national government and not considered as an endogenous variable. Thus, if demand increases, supply will adjust to meet demand as ascertained by national agencies. Per-capita resource consumption is influenced by personal income and the level of accessibility to sources (e.g. Muhammad *et al.*, 2017; Chang, 2014; and Gamage and Jayasena, 2018). Increasing consumption may become a concern for policy makers and other stakeholders in providing sufficient water, energy and food to meet demand. Food consumption diversification especially in staple foods is expected to be implemented to avoid over-dependency on rice locally and nationally. Increasing consumption negatively impacts on the security of all three resources, something to highlight for policy making, especially in the context of meeting multiple SDGs, national targets, as well as environmental protection targets. Increasing food demand also conflicts with the expected reduction in agricultural land.

The decrease in agricultural land on the one hand reduces local food production, impacting negatively on regional food security targets, as shown by this study. However, it also reduces agricultural water demand and improves energy security, positively impacting on progress in these sectors. This demonstrates a choice to be made: if agricultural land loss is beneficial in other sectors, what level of land conversion is acceptable before food production is too adversely impacted? As an example of a benefit, the water supply authority could allocate 'extra' water resulting from agricultural water demand decreases for other uses (e.g. growing domestic demand) to better fulfil demand in other sectors. Such trade-offs can be useful when making decisions related to land conversion in the region, and aligning with medium and long-term planning in other sectors at both local and national level. Results suggest that low levels of agricultural land conversion may be beneficial across nexus sectors (still allowing food production targets to be met and also reducing the stress on water and energy resources), especially if other beneficial measures such as artificial ponds and solar electricity are implemented simultaneously, enhancing synergy amongst policy implementation. These findings support the idea of de Fraiture et al. (2014) on an integrated approach to WEF management. They argue that an integrated approach is necessary not only to ensure sustainability, but could lead to increased economic benefits. This result is surprising because it gives a new limit and a different perspective about agricultural land conversion in this region. The result also indicates that the impact of this policy not only influences the food sector, but also water, energy, and other sectors, issues that were not previously considered locally.

In-migration rate increases due to the indirect implications of regional economic development. With increasing in-migration, there is a need for greater provision of WEF resources. In-migration leads to lowering of the security of all three resources due to increased demand, as shown by model results (Figure 5.10). Distributing the centres of economic growth across the country may reduce excessive urbanization and resource burdens on one region. It is shown (Figure 5.12d and e) that some measures locally such as development of APs and SRE could aid in meeting the increased demand for certain resources without further pressure to the natural resource base.

The nexus approach aims to simultaneously improve the security of water, energy and food by increasing efficiency, reducing trade-offs, exploiting synergies, and improving governance across sectors (Hoff, 2011). From simulated policy scenarios (Figure 5.10 to Figure 5.13) trade-offs and synergy among nexus sectors can be analysed further (summarised in Table 5.6). For instance, the negative impact of moderate land conversion on food production is partially offset by its positive impact for water and energy availability, which can be further enhanced through simultaneous implementation of AP and SRE to increase SSL water and energy (cf. Figure 5.12d and Figure 5.12e). Another example is of the detrimental synergies between industrial growth (Figure 5.9d-f), per-capita consumption (Figure 5.13), and in-migration rate (Figure 5.10). Industrial growth in Karawang triggers an increase in in-migration rate which increases the population, which in turn increases the demand for water, energy and food, especially if per-capita consumption also rises. This leads to reductions in APP and SSL of all resources. These results highlight the need for strong links among institutions related to the three factors (economic growth, migration, consumption) in order to mitigate the negative impacts that these factors lead to, as indicated in this study.

The lack of interaction between industry and agriculture policy may cause agriculture landowners to no longer develop land to produce food, preferring to sell land to be converted into settlements, trade or industrial areas, further reducing food production. New industrial developments could be directed into agro-industries to enhance the value added of agricultural products, increasing the gross reginal domestic product (GRDP) of the agricultural sector. In general, industrial growth leads to detrimental implications for all resources (e.g. Figure 5.9d-f), and conflicts with other objectives, but is beneficial for economic development.

Regarding national mid-term planning targets in water, energy and food, these results are consistent with those of Purwanto *et al.* (2018). It was proposed to conduct strategic options regarding combining industrial and agricultural (agro-industry) activities, where the industrial sector should focus on food processing to absorb existing agricultural products and increase value added for the region. The national government's target is to increase rice production by 26% within the five year planning period. Taking into account the conversion away from agricultural land, the trend of rice production in Karawang showed a decline, conflicting with the national targets and possibly causing issues for food security. However, this scenario reduces agricultural water demand, giving an opportunity to reallocate water and increase domestic access to clean water from 65% to

100% within five years as stated in government planning. The development of artificial ponds could further augment local water supply from 2% to 20% in five years, indicating a nexus wide synergy to be exploited. In the energy sector, the electrification target of 100% can be achieved with almost all the supply coming from PLN. The development of solar rooftop electricity may increase solar energy production from 2% to 19% within five years, far from the national target of a 238% increase, but still an important step towards decarbonisation. Additional efforts to develop solar farm electricity throughout the region instead of only solar rooftop may support the achievement of this target.

5.5 CONCLUSION

The main objective of this research was to analyse quantitatively the implications of planned interventions on APP and SSL of water, energy and food in Karawang Regency. Based on K-WEFS nexus model simulation results, several important results describe the conditions of water, energy and food sectors in the study area including possible actions to be considered. In the simulations, the development of artificial ponds in industrial and settlement areas will increase the SSL of water, but did not significantly affect the SSL food and energy, nor the APP of water, energy and food. The development of solar roof top electricity will only increase APP and SSL of energy. These two policies are expected to be integrated in the medium and long-term development plans by involving related stakeholders so that results are more beneficial for people and ecosystems. Regulation enforcement, subsidy scheme provision, and education for citizens are key to making these policies successful in a sustainable manner. On one hand, agricultural land conversion will lower the SSL food and APP of water, energy and food. However, it will increase the SSL of water and energy. To achieve Regency targets in the Sustainable Food Crops Farmland (LP2B), agricultural conversion rate until 2030 should be less than 1% per year, which can be achieved by enforcing the law and regulations on permits and regional planning. Agro-industrial development is an issue to stimulate a balanced growth between those two sectors. Coordination with irrigation water supply authorities can reduce the trade-offs and improve the accuracy of water allocation.

Trade-offs will always be present in every process of policy making, but reducing the trade-offs and building synergies among institutions and stakeholders will improve the ability of policy makers to take advantage the positive sides and reduce the negative impacts of one or more policies. Policy-makers and other stakeholders can make use this framework, including information generated from this model as one of considerations in the process of evaluation, planning and decision-making related to the supply, demand and security of water, energy and food in the region.

6

SYNTHESIS AND RECOMMENDATION

This chapter comprises general conclusions, main contributions, further research suggestions, and practical recommendations for policy-makers and other stakeholders taken from the whole stage of analysis in this study. In addition, all the results and findings related to the existing water, energy, and food (WEF) nexus studies and frameworks, the economic perspective of WEF sectors, participatory model building, qualitative model development, and quantitative simulation are synthesized and broadly discussed in this part.

6.1 GENERAL CONCLUSIONS

The major objectives of this research is to grasp the WEF security nexus in the local context and evaluate the implications of planned local scale interventions in WEF sectors by developing a qualitative and quantitative analysis framework together with local stakeholders. In accordance with the above main objectives and also five specific objectives of this research, it is now possible to state several general conclusions.

WEF sectors are closely interrelated in any levels and scales, including at the local level and scale as can be found through some analysis in this study. Its complexity is even more challenging to be understood and resolved by all related stakeholders due to several reasons among others the lack of local WEF-related datasets, the lack of capacity of local policy- and decision-makers in grasping the complexity of problems and trade-offs in the WEF sectors, the silo mentality in many WEF-related institutions which causes a lack of coordination in managing WEF resources, and overlapping authority and regulations related to WEF resource management. This study addresses these knowledge and practical gaps via several stages of analysis using various methods such as economic base analysis, stakeholder participation, group model building, and system dynamics modelling. The findings are expected to improve understanding and raise awareness regarding WEF security nexus, including assisting local planners and decision-makers in managing WEF resources.

From economic perspective, this PhD work also investigates the agglomeration and competitiveness of WEF-related sectors and other sectors in three different characteristic local regions in Indonesia and WEF-related sub-sectors in an agriculture–manufacture based region. Using the combination of Location Quotient (LQ) techniques and competitive position (CP) charts, a first assessment of the current levels of growth and agglomeration in WEF sectors based on the gross regional domestic product (GRDP) in the year 2011–2015 and 2000–2015 has been done to bring a better understanding and more comprehensive insights for the planners and policy-makers. Some possible strategies for future sustainable development regarding WEF-related sectors are also formulated and expected to be considered in the future local planning and be employed in conducting preliminary evaluation, particularly the availability of WEF resources to meet local and national WEF security targets.

Stakeholder engagement is strongly suggested in many WEF nexus-related studies and is crucially needed to manage WEF resources, although it is also critical to prevent the delays and slowness in the decision-making processes that may be caused by ineffectiveness of time allocation to accommodate the various kinds of stakeholder's interests. This study demonstrates the effectiveness of qualitative causal loop diagram (CLD) of the WEF nexus in a group model building (GMB) setting in decomposing and understanding a complex system by involving all related stakeholders and the involvement of multi-disciplinary group of researcher, planner and decision-makers during its establishment. It also shows the added value of a qualitative approach, without recourse to computational modelling. The developed qualitative K-WEFS model serves an extensive and coherent overview of the complex WEF systems in the study area that can also be used as a map to clearly depict the positions, tasks and authorities that should be done by each stakeholder to avoid overlapping.

The implications of planned local scale interventions in WEF sectors on the WEF availability per person (APP) and self-sufficiency level (SSL) in the study area are investigated through the developed quantitative K-WEFS model (stock-flow diagram). Trade-offs will always be present in every process of decision-making especially in agricultural sector, land use, and water supply, and industrial development. Reducing the trade-offs by building synergies among institutions and stakeholders will improve the ability to prevent unexpected impacts. The improvements of WEF security in a region can be achieved through some measures among others increase resource supply and efficiency, reduce trade-offs among sectors, establish synergy among related institutions, and improve WEF resource governance.

6.2 SYNTHESIS

6.2.1 Increasing resource supply and efficiency

Based on the results of the quantitative analysis using the K-WEFS model and simulations (Purwanto *et al.*, 2020a; see Chapter 5), one important point is regarding increasing the production and supply of water, energy, and food resources in particularly from local sources. Enhancing the supply of WEF resources locally will strengthen the self-sufficiency of these three resources in the region. Simultaneously, it will also reduce significantly the dependency of WEF resources originating from outside the region or even outside the country (imported resources). In addition, the ability of each local government to ensure its local resources production will give advantage to the national government in achieving WEF security and finally WEF self-sufficiency.

Resources exploitation without balanced and proper management which considers other related factors and sectors is the main challenge and may cause resource insecurity in Indonesia, including in this region. For instance, the agricultural land conversion will decrease rice production but also reduce the amount of irrigated water demand. If planners and policy-makers are able to identify the amount of this reduction, the water supply reallocation can be conducted to meet other water demands such as industry, domestic and other purposes. Another aspect that needs to be considered is the environmental condition and carrying capacity of the region to provide resources. Carrying capacity issues are not specifically discussed in this study. However, ecosystem services as part of a given environment are partly discussed in Chapter 5 although a lack of data prohibited such a comprehensive quantification.

a. Water supply

Several strategies and techniques are available to increase the supply of water from local sources, particularly to balance water availability during the rainy and dry seasons. Such strategies have been discussed in several studies, and include rainwater harvesting, artificial ponds, infiltration wells, groundwater recharging, dam construction, desalination, etc. (Hoff, 2011; ADB, 2016). These techniques, all of which are applicable in Karawang, will provide a significant positive impact if their efficiency can be improved and their contribution maximised.

In this study, scenario analysis shows that artificial pond development a viable option to increase local water supply, especially by considering its relationship with agricultural land conversion, energy generation and food production. However, the scenario results reveal that implementation of artificial ponds will not give a notable impact to the availability per person or the self-sufficiency level of other resources except the water resource itself. Despite this, it may still represent a way for Karawang to be more self-sufficient in terms of water by better exploiting the natural resource potential.

This study suggests several measures to increase the development of artificial ponds in the region, such as: (1) to include in a clear way artificial pond development plans and targets in the mid and long-term planning; (2) increase the developmental growth of artificial ponds especially in the housing, industrial and agricultural areas within the region; (3) to ensure law enforcement and supervision of pond development by industries, housing developers, and other public facilities development and; (4) to extend the purposes of pond development to include flood mitigation, leisure or tourism facilities, fishery potential and other activities, so stakeholders like private companies and communities can be involved to develop, protect and maintain the use of ponds in Karawang.

These steps can be applied in other strategies to increase water supply, so that the acceleration of enhancement of water availability can be done in a better and sustainable way.

b. Energy generation

Energy generation, especially electricity in this region, does not experience such big problems in quantity, because the amount of power produced by the state electricity company (PLN) is expected to be sufficient and even exceed the existing needs. However, with the continuous development of the industrial and trade sectors and the concomitant demand increase, electricity generation must also be the concern of all stakeholders, especially local government. Apart from that, the quality factor due to the low contribution of renewable energy in the local and national energy mix must be a priority in the future. Options for considerable renewable power generation should be considered.

This research addresses this quality issue in electricity generation by considering solar rooftop electricity (SRE) to improve both the quantity and quality (i.e. non-fossil based) of electricity generation. In general, based on simulations, these options do not have a significant effect on other resources (water, land/food). However, by looking at the existing potential and demand, these local and integrated steps related to SRE are expected to make a positive contribution at the national level.

Some possible actions to be taken to optimize the SRE development strategy include: (1) Provide clear rules on SRE development & investment; (2) Educate people on the potential and importance of SRE; (3) Prepare a subsidy scheme for SRE; (4) Collaborate with PLN to increase the growth of SRE; (5) Involve other stakeholders (not only government institutions).

c. Food production

Staple food production, particularly rice, does not experience major constraints in Karawang Regency due to the large of rice harvest area and sufficient irrigation support. However, the continued reduction of agricultural land due to conversion to industrial and residential areas and also the increasing demand for rice must be a serious concern of all stakeholders. In addition, the need for non-staple foods such as meats, fish, vegetables, fruits and others that still relies on the support from outside the region should also be considered with an integrated solution. One of the examples is collaborative planning and action among institutions such as Agriculture Agency, Fishery & Livestock Agency, and Planning Agency within the region and also with other local regions to improve production of non-staple foods.

Increasing food production and diversifying food production and consumption are the measures that can be taken simultaneously to prevent the food self-sufficiency and food security in the region. In addition, maintaining an agriculture land conversion rate of less than 1% per annual until 2030 by improving the process of development planning and supervision will ensure that food production can meet the increasing food demand due to population growth.

6.2.2 Reducing trade-offs

Trade-offs often appear in policy-making processes where they cannot be avoided. Reducing trade-offs and building synergies amongst institutions, policies, and stakeholders will improve the ability of policy makers to take advantage the positive sides and reduce the negative impacts of one or more policies.

a. Agricultural land conversion into industries and settlements

Agricultural land conversion, especially irrigated paddy fields on one hand will decrease rice production overtime. However, it will also reduce the amount of irrigated water demand as irrigated water is only supplied for paddy fields in this region. If planners and policy makers are able to identify the amount of this reduction, the water supply reallocation can be conducted to meet other water demand such as industry, domestic and other purposes. Continuous synergy and communication between PT. PJT II as water supply authority under the national government and also Karawang Regency government, especially Agriculture Agency will greatly help improve the efficiency of the water supply allocated to irrigated agriculture and other purposes within the region.

The implementation of artificial ponds development policies in industrial and residential areas will not only play an important role in increasing the amount of local water production (especially for non-potable uses), but also may reduce flooding events, increase blue open spaces for recreation and biodiversity, recreational facilities and other benefits. The growth of AP development is expected to be even higher along with agricultural land other land uses conversion into industrial and housing areas.

b. Agriculture and industry

The interdependence between these two sectors can be clearly seen in any level of economic development either in positive or negative ways in this region. The potential trade-offs between agricultural and industrial developments are among others agricultural land conversion, the increase of WEF resource consumption, the decrease of irrigation water demand, and the increase of water for industrial purposes. Ideally, agriculture serves industrial sectors in raw material provision, while the industry assist agricultural sector in improving productivity, supplying inputs and technology (Rastegari *et al.*, 2000). Livingstone (1968) argues the importance of 'balanced growth' between agriculture and industry development to obtain the optimal benefits of food provision and labour absorption among others. The complex and dynamic situation in the region need to be understood clearly to ensure the achievement of the development targets sustainably (Rastegari *et al.*, 2000). The ability in analysing the agglomeration and growth of those major sectors in local economic development will bring advantages for policy maker and planner including water-related operators to determine the most appropriate strategies.

Figure 6.1 shows the gross regional domestic product (GRDP) value of agriculture and industrial sectors in two different periods (year 2000-2010 and 2011-2017) in Karawang Regency. It can be clearly seen that there is a significant growth of industrial sector compare to agricultural sector that relatively stagnant in growth. The GRDP growth trends show an increasing gap among these sectors year by year from 2000 to 2017. It may indicates that the trade-offs cannot be handled properly.

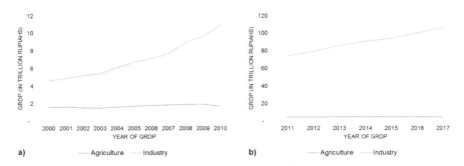

Figure 6.1. The trends of gross regional domestic product (GRDP) of Karawang Regency (a) Year 2000-2010 based on constant price 2000 and (b) Year 2011-2017 based on constant price 2010

The competitive position of the agricultural sector both in the period of 2000-2010 and 2011-2017 were located in quadrant IV (disadvantaged cluster) with LQ value of 0.86-0.45 (non-basic) and P values of -0.44 and -0.14 (see Chapter 3). This cluster indicates the sectors with a low concentration, lack of competitiveness and tends to decline in the sense of sector agglomeration level or concentration. On the other hand, the position of industrial sectors in the first and second periods were in quadrant I (advantaged cluster) with the LQ value of 1.09-1.63 (basic) and P values of 0.03 and 0.47. The sectors located in this quadrant are strong, advanced and expected to become more dominant in the future (*Figure 6.2*).

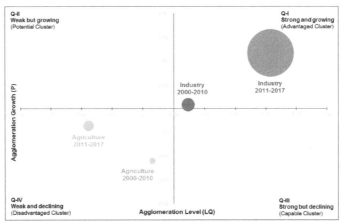

Figure 6.2. The shifting of competitive position of agricultural and industrial sectors in Karawang Regency in the period of 2000-2010 and 2011-2017

The agricultural sector in Karawang Regency is becoming less competitive compared with the industrial sector. Several practical measures that can be considered by local decision makers are, among others: turnaround/retrenchment by changing the development targets and; conglomerate diversification by diversifying the products either agricultural or industrial products in order to support each other and enhance competitiveness. In contrast, the industrial sectors that are located in quadrant I can be treated differently using innovation, vertical integration and concentrated growth approaches (Pearce and Robinson, 2001; see also Chapter 3). Some practical implementation options such as firm acquisition by local owned enterprises, budget allocation to develop basic sources, market expansion and product innovation can be considered to be applied.

The trade-offs between industrial and agricultural development in this region needs to be addressed carefully to avoid mismanagement of WEF resources and to maintain agricultural products sufficiency for local citizens while also offering jobs and economic development in the industrial sector. Integrated planning and appropriate measures are required to achieve the security of WEF sectors. One of recommendations proposed to reduce trade-offs by implementing agricultural-based industry is to balance the growth between these two main sectors in the coming future.

6.2.3 Building institutional and intergovernmental synergies

The achievement of the WEF security targets is still lacking in Karawang due to the large range of authority and responsibility overlapping among institutions at the district, provincial and national levels. Synergy and collaboration are keywords that should be a common concern, so that improvements in the WEF sector can be carried out continuously in a better way. Such a simple collaboration can be seen in the process of WEF security nexus model development by involving all related stakeholders within the region (see Purwanto *et al.*, 2019). Similar things can also be conducted in every stage of planning, implementation, supervision and continuous improvement of each program related to the WEF sectors, so that the achievement of the WEF security targets in the region can be more effectively achieved through better collaboration.

a. Institutional synergies

Currently, collaborative planning and actions to achieve the targets of WEF security in the local level are still very limited. Indeed, for food security, there are already institutions formed in almost all local governments, both province and regency or city who are responsible to achieve food availability, accessibility and utilization. This is in accordance with Law Number 18 of 2012 concerning Food and Law Number 23 of 2014 concerning Local Government Affairs with several amendment and derivative regulations. However, there is no specific authority at the local level to deal with the achievement of water and energy security targets, creating a mismatch in terms of goals, targets, and scales.

The Food Security Agency in a local government is still relatively new and has not been able to become a coordinating body to carry out such collaborative planning and action in this single sector. In addition, unclear responsibilities and roles with other agencies that are also related to food such as the Agriculture Agency, Fisheries Agency, and Animal Husbandry Agency in terms of production and fulfilment of food demand may cause ineffectiveness in synchronizing the food security target achievement for a given region. This complex institutional structure makes meeting targets difficult even within a single sector.

Furthermore, coordinating agencies in water and energy security have not been established yet, even though the challenges in this sector are relatively more complex where a local government has to make such coordination not only with local institution within the region but also other national level institutions. For example, local governments must continue to establish cooperation with the PLN and PERTAMINA as state-owned companies at the national level that provide electricity and fuel to the people in their regions. In addition, the targets and objectives for energy security to be achieved must be synchronized towards a harmony across the whole country. From the water sector, coordination between PJT II as a national water supply authority for irrigation and raw water must be well established so that the aspects of availability, accessibility and quality of water can be achieved sustainably.

The number of institutions and agencies involved in achieving WEF security and overlapping responsibilities and roles as well as the complicated bureaucratic process is about to continue to be an obstacle to the WEF security targets achievements unless some immediate improvements such as collaborative planning and actions, institutional reform, and a harmonized WEF framework are implemented.

The K-WEFS model, both the qualitative causal loop diagram and the quantitative system dynamics model, is expected to contribute towards a turning point for stakeholders, especially at the local level, to be able to see this complex problem in a more integrated manner and better manage it as such. It may reduce the sectoral interests and silo mentality of each institution in planning, implementing and evaluating the WEF security in local regions. In addition, this study is expected to be considered by policy makers, both in the local level and national level, to carry out a comprehensive institutional reform in the WEF sectors. This institutional reform is made for a better integration between sectors and their targets, more effective nexus policy making, and considering other sectoral targets and requirements.

b. Intergovernmental synergies

Each local region has different water, energy, and food resource potentials and demands which should be harnessed and 'optimised' for local, regional, and national resource 'optimisation' across the nexus. For instance, Karawang Regency has a high rice

115

production level, while the level of rice consumption is not too big. On the other hand, Jakarta as the capital city of Indonesia has a high level of rice demand, while there is almost no local rice production. To deal with this issue, the two local regions may carry out continuous cooperation by implementing collaborative planning and action in rice procurement with certain terms and conditions. Each region is expecting to analyse the potential and needs of water, energy, and food resources and then to collaborate with each other in reaching sufficient availability, access, and quality of the resources.

If this kind of scheme continues, the local production systems of water, energy, and food will be stronger and may cause other economic sectors such as transportation, trade and services, and small enterprises to be improved significantly. Meanwhile, the WEF security targets may also be simultaneously achieved to a greater extent. In addition, the increase in the local production systems will reduce the dependence on imports of food, water, and energy at the national level.

The K-WEFS model with several adjustments for other regions is expected to be an example of an instrument that can be employed by local stakeholders and policymakers as a basis planning to evaluate the potential and needs of water, energy, and food resources, so that WEF security or even WEF self-sufficiency can be reached gradually, in line with the development of other economic sectors within the region. The K-WEFS model could therefore be refined and rolled out in other Regencies in Indonesia to support integrated nexus resources decision making.

6.3 MAIN CONTRIBUTIONS

This study provides two major and other specific contributions in the scientific knowledge perspective and societal and practical aspects to support sustainable development. Those two main contributions and other specific contributions are compiled in Table 6.1.

Table 6.1. Scientific and societal contributions of the research

Contribution	Reference
A. Scientific knowledge	
1. The compilation of critiques on the concept, application, and implication of the WEF nexus	Chapter 2 (Table 2.3)
2. The four principles and one perspective for the future WEF nexus frameworks and studies	Chapter 2 (Figure 2.2)
3. The composite methods of location quotient (LQ) and competitive position (CP) using GRDP data	Chapter 3
4. The process of qualitative and quantitative K-WEFS models development in the local context	Chapter 4 & 5
B. Societal and practical implementation	

	Contribution	Reference
1.	Possible strategies for WEF-related sectors for all related stakeholders	Chapter 3 (Table 3.6)
2.	System dynamics group model building script; participatory approach	Chapter 4 (Table 4.2)
3.	The K-WEFS nexus causal loop diagram (CLD)	Chapter 4 (Figure 4.7)
4.	The K-WEFS nexus stock-flow diagram (SFD)	Chapter 5 (Figure 5.5)
5.	The K-WEFS nexus framework for WEF resource evaluation and planning	Chapter 6 (Figure 6.3)

Evaluation and planning are two important measures in resource management to support the achievement of WEF security targets in a region. The entire series of research carried out in this work has provided adequate results that can become an important instrument of evaluation and planning of WEF resources in Karawang Regency. *Figure 6.3* shows several stages of assessment that can be conducted to ensure the effectiveness of WEF resources evaluation and planning process by involving related stakeholders together to minimize unexpected impacts and also to reduce silo mentality within each institution. The principle of "limits to growth" provides different perspective for the policymakers in reducing trade-offs and building synergies among institutions in any horizontal and vertical levels.

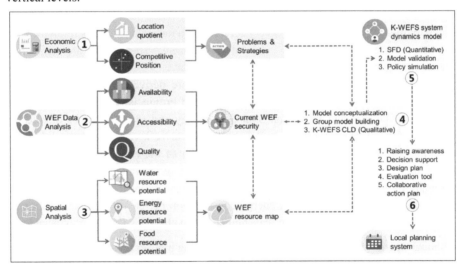

Figure 6.3. The proposed K-WEFS nexus framework for WEF resource evaluation and planning

There are six main stages in this framework (*Figure 6.3*) including; (1) economic base analysis, (2) WEF resource data analysis, (3) spatial WEF resource analysis, (4) qualitative model development, (5) quantitative system dynamics model development, and (6) model implementation and implication in local planning system.

One common method in economic base analysis is the Location Quotient (LQ) (Wang and Hofe, 2007). By analysing production-based gross domestic regional products (GRDP) of this region and West Java Province using LQ method and in combination with Competitive Position (CP) analysis, the basic and non-basic sectors have been identified. Under a general assumption, basic sectors produces more goods or services than the local area needs, so the surplus is possible to be exported to other regions. On the contrary, the non-basic sector is assumed under the level of self-sufficiency, with imports being implied. Additionally, it showed that the general characteristics of WEF-related sectors in a region can be distinguished clearly based on its main economic development focus, so that several possible strategies can be formulated. This preliminary analysis is expected to assist policy-makers, planners and other local stakeholders in doing evaluation and planning of WEF-related sectors in their regions (see Chapter 3), perhaps focussing on those that are economically most important, or instead perhaps focussing efforts on those sectors performing less well to improve their contribution.

The availability of WEF-related datasets at the local level is limited and the data are separated from one another (i.e. not in the same database, another case of a silo-approach). One of data sources that can be collected are annual statistical data issued by Karawang Statistics Agency, which is a combination of data from various agencies in the regions. However, for some specific datasets related to WEF security (i.e. availability, accessibility and quality), we need to collect it through the technical agencies in the local regions, provinces and even the central government. These data are sometimes inconsistent and disconnected each other due to the data updates are not sequential from year to year. For example, data on paddy fields are derived from several agencies including the Agriculture Agency at the District and Provincial levels, the Ministry of Agriculture, District and Provincial Statistics Agency and several sources of research articles. With all these limitations, it is necessary to re-investigate the validity and reliability of the data, including direct confirmation of the related agencies as to the reliability of the data. For some data that is completely unavailable at the local level, assumptions were used based on the data at the provincial, national and even international levels such as rice water demand, energy consumption per capita, and per capita non-staple food consumption in order to scale higher-level values down to the provincial level.

Although spatial WEF resource balance analysis using input-output in combination with GIS-based analysis was not carried out due to the time limitation, this approach is beneficial for local policymakers and planner in determining the WEF resource potentials and future possibilities through maps. For instance, land suitability analysis for food diversification, water production and solar electricity can be calculated and depicted in a map using GIS-based techniques. One preliminary work on the topic of spatial WEF security nexus has been done by Mushayagwazvo (2020). The visualisation method demonstrated showed that it is sometimes easier to understand data in map form compared with models and calculations. The results will bring new perspectives and options to be taken by local stakeholders to improve WEF security in their regions.

Stages 4 and 5 regarding the development of qualitative and quantitative system dynamics models are clearly described in Chapter 4 and Chapter 5. Through a group model building (GMB), a qualitative causal model of a water, energy, and food (WEF) security nexus has been developed. GMB provides the chance to increase problem understanding, raising consensus, and building acceptance and commitment among participants. After dealing with the issues regarding WEF sectors, deriving related variables and eliciting opinions about the interactions, the next stage was to build a conceptual framework to describe the nexus and to develop an integrated causal loop diagram (CLD). In this way, a first step towards breaking silo thinking in local planning, without the need for complex model development, may be attained (Purwanto *et al.*, 2019).

The next step is the development of a stock-flow diagram (SFD) as quantitative model referring to the qualitative CLD that has been established beforehand (see pap 5) and using it as a guide in developing the quantitative model. In this stage the process of model validation, sensitivity and policy analysis, and model simulation are included to ensure that the model is able to represent the current condition and any possible situations due to the planned interventions in WEF related sectors. The incorporation of all stages in this framework to the local planning system is actually the main goal. It is expected that the change in mind-set and the way of thinking from silo mentality to integrated and nexus thinking can help accelerate the achievement of WEF security targets in local levels and finally influence positively the achievement in the national level.

To summarize, there is added value of this whole study that could be practically implemented in the local planning and policymaking of WEF resources in Karawang. The Local Planning Agency (BAPPEDA) may start to involve all WEF-related stakeholders to design and determine practical rules and regulations by referring to the steps above as a guidance. Furthermore, local policy makers together with all stakeholders may propose to the National Government the establishment of a Coordinating Agency of WEF resource security to replace the Food Security Agency. A digital decision support tool of WEF security could be developed to ensure all the WEF-related data from all institutions are integrated and interconnected each other and analysed as a whole, rather than in disconnected silos.

6.4 RECOMMENDATIONS

6.4.1 Recommendation for further scientific research

1. Further WEF nexus-related studies and framework developments are still need to consider four main principles (i.e., to make them more understandable, to ensure

reliable and valid data, to make them adaptable to many diverse situations, and to be applicable across scales) to increase the benefits and improve the role of WEF nexus concept in influencing policy-making and resource planning processes, especially in grounding the concept from the lowest level of government with the perspective of "from local to global".

2. Some limitations in this study can be further complemented by conducting further research in locally-based and more specific issues such as spatial WEF security nexus analysis, WEF security level analysis, and environmental-based and social-based analysis of WEF security nexus.

6.4.2 Practical recommendation

1. Industrialization and economic development can no longer be stopped, because Karawang Regency has been designated as one of the industrial areas in Indonesia. What must be done is to reduce trade-offs, increase local WEF production, and improve synergy among agencies at the local level and between local government of Karawang, West Java Province and the central government to improve WEF security. This improvement is critical because WEF sectors in this region are handled by many institutions either at local or national level, and therefore there is still a lack of coordination and synergy regarding their management. As a result, unexpected impacts can occur like what is happening at present.

2. Agricultural and Industrial developments are two major concerns to achieve people's welfare in this region as stated in the long-term vision of the Regency. Practically, this vision cannot be achieved unless both of industrial and agricultural development are being synchronized and balanced. Agricultural-based Industries approach need to be considered in the future planning of the region. This balanced growth will bring the optimal benefits of food provision and labour absorption in the equal way.

3. The determination of water, energy and food security targets must be clearly stated, more detail and be incorporated in the local long-term planning (RPJPD) which is then being further elaborated in the local mid-term planning (RPJMD), including the measurable stages to ensure the sustainability of its achievement that in line with the national targets. All the targets should be defined by all WEF-related agencies together, so they can mutually assess synergy, trade-offs, and possibly optimise policies.

4. In regard institutional synergies, Coordinating Agency in Water, Energy, and Food Security in every local government level is needed, instead of only Food Security Agency (BKP). This is to assist the government in collaborative planning and actions and synchronizing the target achievements between national targets and local government targets in WEF security. In addition, intergovernmental cooperation

between one and other local regions through the Coordinating Agency can help the local region in fulfilling the WEF demands in accordance with each region's potentials.

5. Breaking the silo thinking is the first measure to make such collaborative actions plans can be implemented by involving not only governmental institutions but also private companies and communities to achieve the targets in WEF security. The use of synchronized methods and standards in measuring WEF security, including data sharing, results sharing and communication at every level of government in a nexus approach are highly important. It can be initiated by the national government to develop and disseminate it to all local regions.

6. One of the strategies that the local government can implement to increase the coverage of water supply, especially potable drinking water, is to involve private sectors in providing drinking water supply through the cooperation between the government and the private sectors so called Public-Private Partnership (PPP) scheme. The increase of water coverage with a better quality and an affordable price for the people will encourage them not to use groundwater anymore for domestic uses. The expansion of drinking water supply coverage will bring implications to the costs that must be invested in relation to new installations and operational costs. However, with this collaboration, it is expected that the target of increasing water security level through the increasing of access and quality of drinking water can be achieved continuously.

7. Many regions in Indonesia have high potential for solar power output. This can be optimized for meeting local energy needs. The development of roof-top solar electricity is expected to increase the availability of local energy source, and may offset fossil-fuel dependant sources to some degree. The industrial development growth may promote the development of residential units in this region, and eventually could bring a positive impact to the development of rooftop solar electricity in Karawang Regency. The role of policymakers in preparing a clear regulations, incentives, subsidies and other measure will encourage people to use this kind of renewable energy.

8. Agriculture land conversion rate, until 2030, should be kept less than 1% per year to achieve the target in the Sustainable Agricultural Land (LP2B). The regulation and law enforcements need to be seriously applied by the Local Government including the supervision of land use permits and agricultural land conversion permit based on local spatial planning.

APPENDIX

INITIAL DATA AND VARIABLE OF K-WEFS STOCK FLOW DIAGRAMS (SFDs)

Variables	Unit	Initial value (2010)	Source and equation
Water			
Total water supply (initial)	m³/y	1,317,081,741	Assumption based on local water production only
Water supply	m³/y	2,487,512,759	local_water_production+imported_water
Local water production	m³/y	1,317,081,741	natural_water_supply+other_water_sources+ AP_water
Other water sources	m³/y	36,528,322	local water company, springs and groundwater sources
AP water (artificial ponds). The initial value is the existing artificial ponds	m³/y	61,508,750	IF TIME = 0 THEN 61508750 ELSE PREVIOUS (SELF, 61508750)+(PREVIOUS(SELF, (housing_area+industrial_area)*potential_AP_water) *AP_growth)
Potential AP water	m³/ha	3,000	AP_area_maximum*AP_depth_maximum
AP depth maximum	m	2	Letter of Minister of Public Work No. 7/2018
AP area maximum	m²	1,500	Letter of Minister of Public Work No. 7/2018
AP growth	%	0.1	Assumption based on the local planning document
Natural water supply	m³	1,521,714,960	10*C_value*annual_rainfall*available_land
Annual rainfall	mm	1,960	Statistics Agency of Karawang (2011)
Imported water	m³/y	1,170,431,018	IF (water_difference<0) THEN ABS (water_difference) ELSE 0
Water difference	m³/y	- 1,170,431,018	total_water_supply-total_water_demand
Total water demand	m³/y	2,343,387,390	Data & calculation
Water demand	m³/y	2,343,387,390	agricultural_water_demand+industrial_water_demand +domestic_water_demand+municipal_water_demand +water_for_energy
Agricultural water demand	m³/y	2,024,112,958	rice_water_demand + livestock_water_demand +fishery_water_demand
Livestock water demand large cattle = 40 ltr/day/animal; small cattle = 6, poultry = 0.6 (SNI 6728.1:2015)	m³/y	6,800,190	IF TIME=0 THEN large_cattle_pop*14.6+small_cattle_pop*2.19+poultry_pop*0.219 ELSE PREVIOUS(SELF, large_cattle_pop*14.6+small_cattle_pop*2.19+poultry_pop*0.219)+ (PREVIOUS (SELF,large_cattle_pop*14.6+small_cattle_pop*2.19 +poultry_pop*0.219)*LS_pop_growth)
Paddy water demand	m³/y	1,988,055,463	paddy_water_req*paddy_production
Paddy water req (requirement)	m³/ton	1,432	irrigated lowland rice production system (www.knowledgebank.irri.org)
Fishery water demand	m³/y	29,257,305	((fishery_water_req/1000)*365)*(aquaculture_area* 1000)

Variables	Unit	Initial value (2010)	Source and equation
Fishery water req (requirement)	mm/d/ha	7	Ministry of Public Work (1993)
Industrial water demand	m³/y	115,416,720	industrial_water_req*industrial_area
Industrial water req (requirement)	m³/ha/y	15,513	18–67 m³/day/ha. Average: 15,513 m³/ha/y (ICWA-ADB, 2016)
Domestic water demand	m³/y	117,028,505	percapita_water_cons*population
Per capita water cons (consumption)	m³/cap	55	(SNI 6728.1:2015)
Municipal water demand	m³/y	23,405,701	15-30% of total domestic consumption (SNI 6728.1:2015). 30% x 117,028,505 = 23,405,701 m³/y. Actual: 79,416,927 m³/y.
APP water (Availability per person)	m³/cap	1,101	water_supply/population
SSL water (Self-sufficiency level)	dmnl	0.56	local_water_production/water_demand
Water security	dmnl	2.38	(household_WS)+(economic_WS)+(urban_WS)+(environmental_WS)+(water_related_disaster_resilience) *) *maximum value is 5*
Environmental WS	dmnl	0.1	WI12*0.1
Water quality index (WQI)	dmnl	20	Analysis (calculation based on Ministry of Environment guideline by considering 8 river water parameters i.e. TSS, DO, BOD, COD, total phosphate, faecal coli and total coliform). **Parameter (avg)** TSS 67.5 mg/l, **Standard class I** 50 mg/l; DO 1.9 mg/l, 6 mg/l; BOD 12.2 mg/l, 2 mg/l; COD 96 mg/l, 10 mg/l; PO4 0.8 mg/l, 0.2 mg/l; Faecal coli 76500/100ml, 100/100ml; Total coliform 179000/100ml, 1000/100ml. Average pollution index = 192.7 and >6.88 $\rightarrow$ WQI = 20
Urban WS	dmnl	0.1	WI11*0.1
PDAM municipal coverage	%	9.59	Analysis (based on municipal coverage of piped water)
Household WS	dmnl	0.45	(WI1*0.1)+(WI2*0.05)+(WI3*0.05)
PDAM household coverage	%	11.43	Analysis (based on household coverage of piped water). By 2019, the coverage is was reaching 23% of total population
Use of septic tank	%	67.35	Welfare Statistics of Karawang, BPS
Safe drinking water source	%	44.4	Safe drinking water sources consist of piped water (PDAM), water pumps, protected wells or springs and rain water (not including bottled water and refill water). *Welfare Statistics of Karawang, BPS
Without safe drinking water	%	55.6	1-safe_drinking_water_source
Economic WS	dmnl	1.229	(broad_economy_WS*0.4)+(agriculture_WS*0.3)+(energy_WS*0.15)+(industrial_WS*0.15)
Agriculture WS	dmnl	0.75	(WI7*0.045)+(WI8*0.105)
Industrial WS	dmnl	5	WI10*1
Energy WS	dmnl	1	WI9*1
Broad economy WS	dmnl	0.26	(WI4*0.06)+(WI5*0.06)+(WI6*0.08)

Variables	Unit	Initial value (2010)	Source and equation
Water supply reliability	dmnl	0.675	Based on average coefficient of monthly rainfall variation in Java Island (Hatmoko et al. 2017)
Ratio of storage to TWS	dmnl	0.026	AP_water/water_supply
Water use pressure	dmnl	0.999	water_demand/water_supply
Water related disaster resilience	dmnl	0.5	WI13*0.1
High risk floods area	ha	128.98	BNPB-RI (using flood disaster risk index)
Water for energy	m³/y	51,720,655	energy_supply*(energy_water_intensity/1000)
Energy water intensity	m³/MWh	13	Indonesia Country Water Assessment-ADB (2016)
WI1 (water indicator 1) Percentage of PDAM household service coverage PDAM (%)	dmnl	1	IF PDAM_household_coverage>=0.8 THEN 5 ELSE; IF PDAM_household_coverage<0.8 AND PDAM_household_coverage>=0.6 THEN 4 ELSE; IF PDAM_household_coverage<0.6 AND PDAM_household_coverage>=0.4 THEN 3 ELSE; IF PDAM_household_coverage<0.4 AND PDAM_household_coverage >=0.2 THEN 2 ELSE 1
WI2 (water indicator 2) Percentage of the use of septic tank (%)	dmnl	4	IF use_of_septic_tank>=0.8 THEN 5 ELSE; IF use_of_septic_tank<0.8 AND use_of_septic_tank>=0.6 THEN 4 ELSE; IF use_of_septic_tank<0.6 AND use_of_septic_tank>=0.4 THEN 3 ELSE; IF use_of_septic_tank>0.4 AND use_of_septic_tank>=0.2 THEN 2 ELSE 1
WI3 (water indicator 3) Percentage of safe drinking water source (%)	dmnl	3	IF safe_drinking_water_source>=0.8 THEN 5 ELSE; IF safe_drinking_water_source<0.8 AND safe_drinking_water_source>=0.6 THEN 4 ELSE; IF safe_drinking_water_source<0.6 AND safe_drinking_water_source>=0.4 THEN 3 ELSE; IF safe_drinking_water_source<0.4 AND safe_drinking_water_source>=0.2 THEN 2 ELSE 1
WI4 (water indicator 4) Water supply reliability (%)	dmnl	2	IF water_supply_reliability<=0.2 THEN 5 ELSE; IF water_supply_reliability>0.2 AND water_supply_reliability<=0.4 THEN 4 ELSE; IF water_supply_reliability>0.4 AND water_supply_reliability<=0.6 THEN 3 ELSE; IF water_supply_reliability>0.6 AND water_supply_reliability<=0.75 THEN 2 ELSE 1
WI5 (water indicator 5) Water use pressure	dmnl	1	IF water_use_pressure<=0.1 THEN 5 ELSE; IF water_use_pressure>0.1 AND water_use_pressure <=0.2 THEN 4 ELSE; IF water_use_pressure>0.2 AND water_use_pressure <=0.4 THEN 3 ELSE; IF water_use_pressure>0.4 AND water_use_pressure <=0.8 THEN 2 ELSE 1
WI6 (water indicator 6) Ratio of water storage and total water resource	dmnl	1	IF ratio_of_storage_to_TWS>=0.5 THEN 5 ELSE; IF ratio_of_storage_to_TWS<0.5 AND ratio_of_storage_to_TWS>=0.2 THEN 4 ELSE; IF ratio_of_storage_to_TWS<0.2 AND ratio_of_storage_to_TWS>=0.05 THEN 3 ELSE; IF ratio_of_storage_to_TWS<0.05 AND ratio_of_storage_to_TWS>=0.03 THEN 2 ELSE 1
WI7 (water indicator 7) Productivity of	dmnl	5	IF paddy_productivity>=5.5 THEN 5 ELSE; IF paddy_productivity=4.5 AND paddy_productivity

Variables	Unit	Initial value (2010)	Source and equation
irrigated agriculture land (ton/ha)			<5.5 THEN 4 ELSE; IF paddy_productivity=3.5 AND paddy_productivity <4.5 THEN 3 ELSE; IF paddy_productivity=2.5 AND paddy_productivity <3.5 THEN 2 ELSE 1
WI8 (water indicator 8) Rice self-sufficiency (ratio of rice production and rice consumption)	dmnl	5	IF rice_self_sufficiency>=3 THEN 5 ELSE; IF rice_self_sufficiency<3 AND rice_self_sufficiency >=1.5 THEN 4 ELSE IF rice_self_sufficiency<1.5 AND rice_self_sufficiency>=1 THEN 3 ELSE; IF rice_self_sufficiency<1 AND rice_self_sufficiency>=0.5 THEN 2 ELSE 1
WI9 (water indicator 9) Installed capacity of hydropower, micro HP and mini HP (MW)	dmnl	1	IF hydropower_IC>=1000 THEN 5 ELSE; IF hydropower_IC<1000 AND hydropower_IC>=500 THEN 4 ELSE IF hydropower_IC<500 AND hydropower_IC>=250 THEN 3 ELSE; IF hydropower_IC<250 AND hydropower_IC>=100 THEN 2 ELSE 1
WI10 (water indicator 10) Industrial water (10^3 m^3)	dmnl	5	IF industrial_water_demand>=30000000 THEN 5 ELSE; IF industrial_water_demand>=10000000 AND industrial_water_demand <30000000 THEN 4 ELSE; IF industrial_water_demand>=5000000 AND industrial_water_demand <10000000 THEN 3 ELSE; IF industrial_water_demand>=3000000 AND industrial_water_demand <5000000 THEN 2 ELSE 1
WI11 (water indicator 11) Percentage of PDAM* municipal service coverage (%) *PDAM = Local Water Company	dmnl	1	IF PDAM_municipal_coverage>=0.80 THEN 5 ELSE; IF PDAM_municipal_coverage<0.8 AND PDAM_municipal_coverage >=0.6 THEN 4 ELSE IF PDAM_municipal_coverage<0.6 AND PDAM_municipal_coverage >=0.4 THEN 3 ELSE IF PDAM_municipal_coverage<0.40 AND PDAM_municipal_coverage >=0.20 THEN 2 ELSE 1
WI12 (water indicator 12) Water quality index	dmnl	1	IF water_quality_index>=80 AND water_quality_index <90 THEN 5 ELSE; IF water_quality_index>=70 AND water_quality_index <80 THEN 4 ELSE; IF water_quality_index>=60 AND water_quality_index <70 THEN 3 ELSE; IF water_quality_index>=50 AND water_quality_index <60 THEN 2 ELSE 1
WI13 (water indicator 13) High-risk floods area (km²)	dmnl	5	IF high_risk_floods_area<=400 THEN 5 ELSE; IF high_risk_floods_area>400 AND high_risk_floods_area <=800 THEN 4 ELSE; IF high_risk_floods_area>800 AND high_risk_floods_area<=1200 THEN 3 ELSE; IF high_risk_floods_area>1200 AND high_risk_floods_area <=1600 THEN 2 ELSE 1
Energy			
Total energy supply (initial) Local production	kWh/y	216,500,129	3,275,308,312 (imported), 3,481,263,599 (total) Statistics Agency of West Java (2011) & PT. PLN West Java & Banten Distribution
Energy supply	kWh/y	3,481,263,599	local_energy_production+imported_energy
Local energy production	kWh/y	216,500,129	other_energy_sources+solar_energy
Other energy source	kWh/y	181,460,440	Coal power plant in industry and other small diesel power plants

126

Variables	Unit	Initial value (2010)	Source and equation
Solar energy	kWh/y	35,039,689	IF TIME = 0 THEN 0 ELSE PREVIOUS(SELF, 0) + (PREVIOUS(SELF, ((housing_area*Phousing*11.5*365)+(industrial_area *Pindustry*11.5*365)))*solar_growth)
Phousing (Potential power generated from housing area)	kW/ha	216	This value is based on assumption and calculation: 30% open space, 70% built space regulation, Sun irradiance 5.25 kWh/m^2/day, 11.5 hours (duration/day); 365 (days per year), 18 modules per house, module efficiency 14.2%, Losses of cable 2%, losses inverter 2%, Module size 1.65 (length) x 0.99 (width) m^2
Pindustry (Potential power generated from industrial area)	kW/ha	432	Assumed as 2 x Phousing because the rooftop size and position in industrial area are 2 times compare with housing area
Solar growth	dmnl	0.001	Assumption
Energy difference	kWh/y	-3,762,011,802	total_energy_supply-total_energy_demand
Total energy demand	kWh/y	3,978,511,931	Data & calculation
Energy demand	kWh/y	3,978,511,931	Data & calculation
Domestic energy demand	kWh/y	1,349,019,494	Data & calculation
Per capita energy cons (electricity)	kWh/ cap	634	Ministry of Energy and Resources
Industrial energy demand	kWh/y	3,449,832,624	industrial_area*industrial_energy
Industrial energy	kWh/ha /y	450,000	Electricity demand according to BPPT is 0.15-0.2 MW/ha. By assuming 10 hours per day and 300 days per year, total electricity consumption is assumed as 450,000 kWh/ha/y
Agricultural energy demand	kWh/y	738,857,909	paddy_production*agricultural_energy
Agricultural energy	kWh/ ton	533	http://www.fao.org/3/x8054e/x8054e05.htm
APP energy (Availability per person)	kWh/ cap	1,870	energy_supply/population
SSL energy (self-sufficiency level)	dmnl	0.034	local_energy_production/energy_demand
Energy security	dmnl	5	(energy_accessibility)+(energy_availability)
Energy availability	dmnl	2.5	(EI1*0.5)
Ratio supply and demand	dmnl	0.630	energy_supply/energy_demand
Energy accessibility	dmnl	2.5	(0.25*EI3)+(0.25*EI2)
Access to electricity	%	99.63	Statistics Agency of Karawang
Without electricity	%	0.37	Statistics Agency of Karawang
Access to household fuel	%	90	Statistics Agency of Karawang
Hydropower IC	MW	6.5	PJT II
Energy for water	kWh/ m^3	0.37	(Hoff 2011)
Imported energy	Kwh/y	3,275,308,312	Statistics Agency of West Java (2011) & PT. PLN West Java & Banten Distribution

Variables	Unit	Initial value (2010)	Source and equation
EI1 (energy indicator 1), The ratio of electricity supply and electricity demand	dmnl	5	IF ratio_supply_&_demand>=0.8 THEN 5 ELSE; IF ratio_supply_&_demand<0.8 AND ratio_supply_&_demand >=0.60 THEN 4 ELSE; IF ratio_supply_&_demand<0.60 AND ratio_supply_&_demand>=0.40 THEN 3 ELSE; IF ratio_supply_&_demand<0.40 AND ratio_supply_&_demand>=0.2 THEN 2 ELSE 1
EI2 (energy indicator 2) Access to electricity (% of population with access to electricity)	dmnl	5	IF access_to_electricity>=0.8 THEN 5 ELSE; IF access_to_electricity<0.8 AND access_to_electricity >=0.60 THEN 4 ELSE; IF access_to_electricity<0.60 AND access_to_electricity>=0.40 THEN 3 ELSE; IF access_to_electricity<0.40 AND access_to_electricity>=0.2 THEN 2 ELSE 1
EI3 (energy indicator 3) Access to clean cooking (% of population using LPG/PNG for cooking)	dmnl	5	IF access_to_household_fuel>=0.8 THEN 5 ELSE; IF access_to_household_fuel<0.8 AND access_to_household_fuel>=0.6 THEN 4 ELSE; IF access_to_household_fuel<0.6 AND access_to_household_fuel >=0.4 THEN 3 ELSE; IF access_to_household_fuel<0.4 AND access_to_household_fuel >=0.2 THEN 2 ELSE 1
Food			
Total food supply	ton	854,892	INIT(854,892)
Food supply	ton/y	854,892	local_food_production+imported_food
Local food production	ton/y	854,892	(paddy_production*0.58)+CCS_production+livestock_production+fishery_production
Paddy production	ton/y	1,388,307	agriculture_area*paddy_productivity*cropping_intensity
Paddy productivity	ton/ha	6.550	=graph(time) (2010, 6.550), (2011, 6.842), (2012, 7.079), (2013, 7.140), (2014, 7.219), (2015, 7.254), (2016, 7.272), (2017, 7.272), (2018, 7.246), (2019, 7.228), (2020, 7.211), (2021, 7.228), (2022, 7.263), (2023, 7.263), (2024, 7.289), (2025, 7.289), (2026, 7.263), (2027, 7.263), (2028, 7.263), (2029, 7.237), (2030, 7.211)
Cropping intensity		1.950	=graph(time) (2010, 1.950), (2011, 2.000), (2012, 2.000), (2013, 2.026), (2014, 2.026), (2015, 2.026), (2016, 2.018), (2017, 2.018), (2018, 2.026), (2019, 2.018), (2020, 2.035), (2021, 2.044), (2022, 2.044), (2023, 2.044), (2024, 2.044), (2025, 2.044), (2026, 2.044), (2027, 2.044), (2028, 2.026), (2029, 2.026), (2030, 2.035)
Livestock production	ton/y	12,941	IF TIME = 0 THEN (meat+egg) ELSE PREVIOUS(SELF, (meat+egg)) + (PREVIOUS(SELF, (meat+egg))*livestock_growth)
Meat	ton/y	8,988	Agriculture Agency of Karawang (2016)
Egg	ton/y	4,450	Agriculture Agency of Karawang (2016)
Livestock growth	%	-3.7	Agriculture Agency of Karawang (2016)
Large cattle pop	animal	17,074	Agriculture Agency of Karawang (2016)
Small cattle pop	animal	1,884,460	Agriculture Agency of Karawang (2016)
Poultry pop	animal	10429569	Agriculture Agency of Karawang (2016)
Fishery production	ton/y	56,244	IF TIME = 0 THEN fishery_productivity* aquaculture_area ELSE PREVIOUS(SELF, (fishery_productivity*aquaculture_area)) +

Variables	Unit	Initial value (2010)	Source and equation
			(PREVIOUS(SELF, (fishery_productivity* aquaculture_area))*fishery_growth)
Fishery productivity	ton/ha	2.5	Strategic Planning of Fishery Agency 2016-2021
CCS production	ton/y	13625	IF TIME = 0 THEN (corn_prod+cassava_prod+sweet_potato_prod) ELSE PREVIOUS (SELF, (corn_prod+cassava_prod+sweet_potato_prod)) + (PREVIOUS(SELF, (corn_prod+cassava_prod+sweet_potato_prod))*CCS _growth)
CCS growth	%	-0.1	Statistics Agency & Agriculture Agency of Karawang
Corn prod	ton/y	6,224	Statistics Agency & Agriculture Agency of Karawang
Cassava prod	ton/y	7,061	Statistics Agency & Agriculture Agency of Karawang
Sweet potato prod	ton/y	354	Statistics Agency & Agriculture Agency of Karawang
Total food demand	ton/y	304,337	INIT(304337)
Food demand	ton/y	304,337	domestic_food_cons+non_domestic_food_demand
Non domestic food demand	ton/y	27,766	0.02*paddy_production
Domestic food demand	ton/y	276,613	population*(percapita_staple_food_cons+percapita_non_staple_food_cons)
Percapita non staple food cons	ton/cap /y	0.032	Pola Pangan Harapan of West Java 2018
Percapita staple food cons	ton/cap /y	0.118	Pola Pangan Harapan of West Java 2018
Domestic staple food demand	ton/y	236,184	population*percapita_staple_food_cons
APP food (availability per person)	ton/cap	0.4	food_supply/population
SSL food (self-sufficiency level)	dmnl	2.533	local_food_production/food_demand
Ratio of cons to food avail	dmnl	0.288	domestic_staple_food_demand/(paddy_production*0.58+CCS_production)
Rice self sufficiency	dmnl	0.198	(paddy_production*0.58)/domestic_rice_demand
Domestic rice demand	ton/y	219,162	percapita_rice_cons*population
Per capita rice cons	ton/cap /y	0.105	Pola Pangan Harapan of West Java 2018
Food security	dmnl	3.7	(food_availability)+(food_accessibility)+(food_utilization)
Food availability	dmnl	1.5	0.3*FI1
Food accessibility	dmnl	0.825	(0.15*FI2)+(0.075*FI3)+(0.075*(1-FI4))
Food utilization	dmnl	1.4	(0.05*FI5)+(0.15*FI6)+(0.05*FI7)+(0.05*FI8)+(0.1*FI9)
Food difference	ton/y	550,555	total_food_supply-total_food_demand
Imported food	ton/y	0	IF (food_difference<0) THEN ABS (food_difference) ELSE 0
Total NSF supply	ton/y	36,049	Agriculture Agency of Karawang (2016)
NSF supply (non-staple food)	ton/y	64,190	livestock_production+fishery_production+imported_NSF
Non-staple food difference	ton/y	-4,379	total_NSF_supply-total_NSF_demand

Variables	Unit	Initial value (2010)	Source and equation
Imported NSF	ton/y	4,379	IF (non_staple_food_difference<0) THEN ABS(non_staple_food_difference) ELSE 0
Total NSF demand	ton/y	40,428	Data & calculation
NSF demand	ton/y	68,089	population*percapita_non_staple_food_cons
FI1 (food indicator 1) Ratio of normative consumption per capita to the net food availability	dmnl	5	IF ratio_of_cons_to_food_avail<0.75 THEN 5 ELSE; IF ratio_of_cons_to_food_avail>=0.75 AND ratio_of_cons_to_food_avail<1.0 THEN 4 ELSE; IF ratio_of_cons_to_food_avail>=1 AND ratio_of_cons_to_food_avail<1.25 THEN 3 ELSE; IF ratio_of_cons_to_food_avail>=1.25 AND ratio_of_cons_to_food_avail<1.50 THEN 2 ELSE 1
FI2 (food indicator 2) Percentage of population living below the poverty line	dmnl	5	IF people_below_poverty<0.15 THEN 5 ELSE; IF people_below_poverty>=0.15 AND people_below_poverty<0.20 THEN 4 ELSE; IF people_below_poverty>=0.20 AND people_below_poverty<0.25 THEN 3 ELSE; IF people_below_poverty>=0.25 AND people_below_poverty<0.35 THEN 2 ELSE 1
FI3 (food indicator 3) Percentage of households with a proportion of food expenditure more than 65% of total expenditure	dmnl	5	IF household_with_65%_for_food <0.2 THEN 5 ELSE; IF household_with_65%_for_food >=0.20 AND household_with_65%_for_food <0.30 THEN 4 ELSE; IF household_with_65%_for_food >=0.30 AND household_with_65%_for_food<0.40 THEN 3 ELSE; IF household_with_65%_for_food >=0.40 AND household_with_65%_for_food<0.50 THEN 2 ELSE 1
FI4 (food indicator 4) Percentage of households without access to electricity	dmnl	5	IF without_electricity<0.2 THEN 5 ELSE; IF without_electricity>=0.20AND without_electricity <0.30 THEN 4 ELSE; IF without_electricity>=0.30 AND without_electricity <0.40 THEN 3 ELSE; IF without_electricity>=0.40 AND without_electricity<0.50 THEN 2 ELSE 1
FI5 (food indicator 5) The average length of schooling of women more than 15 years old	dmnl	1	IF women_15_yrs_length_of_schooling>=8.5 THEN 5 ELSE; IF women_15_yrs_length_of_schooling<8.5 AND women_15_yrs_length_of_schooling>=7.5 THEN 4 ELSE; IF women_15_yrs_length_of_schooling<7.5 AND women_15_yrs_length_of_schooling>=6.5 THEN 3 ELSE; IF women_15_yrs_length_of_schooling<6.5 AND women_15_yrs_length_of_schooling>=6 THEN 2 ELSE 1
FI6 (food indicator 6) Percentage of households without access to clean water	dmnl	3	IF without_safe_drinking_water<0.4 THEN 5 ELSE; IF without_safe_drinking_water>=0.4 AND without_safe_drinking_water <0.5 THEN 4 ELSE; IF without_safe_drinking_water>=0.5 AND without_safe_drinking_water<0.6 THEN 3 ELSE; IF without_safe_drinking_water>=0.6 AND without_safe_drinking_water<0.7 THEN 2 ELSE 1
FI7 (food indicator 7) Ratio of total population per health worker to population density	dmnl	5	IF health_worker_pop_density_ratio<10 THEN 5 ELSE; IF health_worker_pop_density_ratio>=10 AND health_worker_pop_density_ratio<15 THEN 4 ELSE; IF health_worker_pop_density_ratio>=15 AND health_worker_pop_density_ratio<20 THEN 3 ELSE; IF health_worker_pop_density_ratio>=20

Variables	Unit	Initial value (2010)	Source and equation
			AND health_worker_pop_density_ratio<30 THEN 2 ELSE 1
FI8 (food indicator 8) Ratio of total population per health worker to population density	dmnl	3	IF prevalence_stunting_toddler<0.2 THEN 5 ELSE; IF prevalence_stunting_toddler>=0.2 AND prevalence_stunting_toddler<0.29 THEN 4 ELSE; IF prevalence_stunting_toddler>=0.30 AND prevalence_stunting_toddler<0.39 THEN 3 ELSE 2
FI9 (food indicator 9) Life expectancy at birth	dmnl	5	IF life_expectancy>67 THEN 5 ELSE; IF life_expectancy>64 AND life_expectancy<=67 THEN 4 ELSE; IF life_expectancy>61 AND life_expectancy<=64 THEN 3 ELSE; IF life_expectancy>58 AND life_expectancy<=61 THEN 2 ELSE 1
Population			
Population	people	2,127,791	Statistics Agency of Karawang
Births	people	34,045	Statistics Agency of Karawang
Birth rate	%	1.6	Health Agency WJ & RPJMD of Karawang
Deaths	people	9,575	Health Agency WJ
Death rate	%	0.45	Health Agency WJ
Life expectancy	year	71.35	Statistics Agency of Karawang
In migration	people	36,172	Statistics of Migration WJ
In migration rate	%	1.7	Statistics of Migration WJ
Outmigration	people	21,277	Statistics of Migration WJ
Outmigration rate	%	1.0	Statistics of Migration WJ
Prevalence stunting toddler	%	34,87	Statistics of Health
Women 15 yrs length of schooling	%	5.27	Statistics Agency of Karawang
Health worker pop density ratio	dmnl	3.27	Statistics Agency of Karawang
People below poverty	%	12.21	Statistics Agency of Karawang
Household with 65% for food	%	10	Statistics Agency of Karawang
Ecosystem services			
Environment quality index	dmnl	40.7	(0.3*water_quality_index)+(0.3*air_quality_index)+(0.4*land_cover_index)
Land cover index	dmnl	44.9	0.23*FCI+0.24*FPI+0.30*LCC+0.15*WBC+0.08*HCI
HCI (habitat condition index)	dmnl	30	Ministry of Environment
WBC (water body conservation)	dmnl	10	Ministry of Environment
FPI (forest performance index)	dmnl	50	Ministry of Environment
FCI (forest cover index)	dmnl	27.69	100-(84.3-((forest_area/available_land)*100)*(50/54.3))
LCC (land cover condition)	dmnl	79.74	(1- C_value*0.625)*100
Air quality index	dmnl	56	Annual Performance Report (LAKIP) of Karawang 2017
C value (Run off coefficient)	dmnl	0.324	((agriculture_area*0.3)+(industrial_area*0.7)+(housing_area*0.5)+(aquaculture_area*0.18)+(forest_area*0.2)+(other_area*0.35))/available_land

131

Variables	Unit	Initial value (2010)	Source and equation
Economic sector			
Total GRDP	Million Rp	99,964,131	Gross Regional Domestic Product (Statistics Agency of Karawang)
GRDP growing	Million Rp	7,443,206	(GRDP_WE+GRDP_agriculture+GRDP_industry+GRDP_others)*GRDP_growth_rate
GRDP growth rate	%	7.47	Statistics Agency of Karawang
GRDP agriculture	Million Rp	4,570,801	Statistics Agency of Karawang
GRDP WE (Water and energy sector)	Million Rp	993,521	Statistics Agency of Karawang
GRDP industry	Million Rp	68,409,475	Statistics Agency of Karawang
GRDP others	Million Rp	25,667,522	Statistics Agency of Karawang
Percapita GRDP	Million Rp	46.82	Data & calculation
Land sector			
Housing area	ha	24,121	Rafiuddin et al. 2016
Housing growing	ha	955.19	housing_area*housing_growth
Housing growth	%	3.96	Spatial Planning of Karawang 2011-2031
Industrial area	ha	7,440	Rafiuddin et al. 2016; RTRW 2011-2031
Industrial growing	ha	294.62	industrial_area*industrial_growth
Industrial growth	%	3.96	Spatial Planning of Karawang 2011-2031
Forest area	ha	7,104	Rafiuddin et al. 2016
Forest reducing	ha	6.39	forest_area*forest_shrink_rate
Forest shrink rate	%	0.09	Rafiuddin et al. 2016
Agriculture area	ha	108,695	Rafiuddin et al. 2016
Agriculture reducing	ha	1,554	agriculture_area*agriculture_land_shrink_rate
Agriculture shrink rate	%	1.43	Rafiuddin et al. 2016
Aquaculture area	ha	18,748	Rafiuddin et al. 2016
Aquaculture growing	ha	103	aquaculture_area*aquaculture_growth
Aquaculture growth rate	%	0.55	Rafiuddin et al. 2016
Other area	ha	25,756	available_land-(housing_area+industrial_area+aquaculture_area+agriculture_area+forest_area)
Available land	ha	191,864	Rafiuddin et al. 2016
Area check	ha	191,864	housing_area+industrial_area+aquaculture_area+agriculture_area+forest_area+other_area

REFERENCES

Ababaei B, Ramezani Etedali H. 2017. Water footprint assessment of main cereals in Iran. *Agricultural Water Management* **179**: 401–411 DOI: 10.1016/j.agwat.2016.07.016

ADB. 2016a. Indonesia Country Water Assessment. Asian Development Bank, Manila. Available at: https://openaccess.adb.org

ADB. 2016b. *Asian Water Development Outlook 2016*. Asian Development Bank (ADB): Manila. Available at: http://www.adb.org/sites/default/files/pub/2013/asian-water-development-outlook-2013.pdf

Ahmad A, Khan S. 2017. Water and Energy Scarcity for Agriculture: Is Irrigation Modernization the Answer? *Irrigation and Drainage* **66** (1): 34–44 DOI: 10.1002/ird.2021

Al-Ansari T, Korre A, Nie Z, Shah N. 2015. Development of a life cycle assessment tool for the assessment of food production systems within the energy, water and food nexus. *Sustainable Production and Consumption* **2** (March): 52–66 DOI: 10.1016/j.spc.2015.07.005

Albrecht TR, Crootof A, Scott CA. 2018. The water-energy-food nexus: A systematic review of methods for nexus assessment. *Environmental Research Letters* **13**: 1–26 DOI: 10.1088/1748-9326/aaa9c6

Alhowaish AK, Alsharikh MA, Alasmail MA, A AZ. 2015. Location quotient technique and economy analysis of regions: Tabuk Province of Saudi Arabia as a case study. *International Journal of Science and Research (IJSR)* **4** (12): 1756–1761

Allende A, Monaghan J. 2015. Irrigation water quality for leafy crops: A perspective of risks and potential solutions. *International Journal of Environmental Research and Public Health* **12** (7): 7457–7477 DOI: 10.3390/ijerph120707457

Allouche J, Middleton C, Gyawal D. 2014. *Nexus Nirvana or Nexus Nullity? A dynamic approach to security and sustainability in the water-energy-food nexus*. Available at: www.steps-centre.org/publications

Altamirano MA, Bodegom AJ van, Linden N van der, Rijke H de, Verhagen J, Bucx T, Boccalon A, Zwaan B van der. 2018. Operationalizing the WEF nexus quantifying the trade-offs and synergies between the water-energy and food sectors.pdf

Andrews-Speed P, Bleischwitz R, Boersma T, Johnson C, Kemp G, VanDeveer SD. 2012. *The global resource nexus: The struggles for land, energy, food, water, and minerals*. The Transatlantic Academy: Washington DC.

Artioli F, Acuto M, McArthur J. 2017. The water-energy-food nexus: An integration agenda and implications for urban governance. *Political Geography* **61**: 215–223 DOI: 10.1016/J.POLGEO.2017.08.009

Bakhshianlamouki E, Masia S, Karimi P, van der Zaag P, Sušnik J. 2020. A system dynamics model to quantify the impacts of restoration measures on the water-energy-food nexus in the Urmia lake Basin, Iran. *Science of The Total Environment* **708**: 134874 DOI: 10.1016/J.SCITOTENV.2019.134874

Bala BK, Alias EF, Arshad FM, Noh KM, Hadi AHA. 2014. Modelling of Food Security in Malaysia. *Simulation Modelling Practice and Theory* **47**: 152–164 DOI: 10.1016/j.simpat.2014.06.001

Ball VE, Färe R, Grosskopf S, Margaritis D. 2015. The role of energy productivity in U.S. agriculture. *Energy Economics* **49**: 460–471 DOI: 10.1016/j.eneco.2015.03.006

Bazilian M, Rogner H, Howells M, Hermann S, Arent D, Gielen D, Steduto P, Mueller A, Komor P, Tol RSJ, et al. 2011. Considering the Energy, Water and Food Nexus: Towards an Integrated Modelling Approach. *Energy Policy* **39** (12): 7896–7906 DOI: 10.1016/j.enpol.2011.09.039

Bell A, Matthews N, Zhang W. 2016. Opportunities for Improved Promotion of Ecosystem Services in Agriculture Under the Water-Energy-Food Nexus. *Journal of Environmental Studies and Sciences* **6** (1): 183–191 DOI: 10.1007/s13412-016-0366-9

Bellfield H, Leggett M, Trivedi M, Pareira J, Gangga A. 2016. How can Indonesia achieve water, energy and food security without eroding its natural capital? WCS Indonesia and Global Canopy Programme, Indonesia. Available at: www.globalcanopy.org

Benson D, Gain AK, Rouillard JJ. 2015. Water governance in a comparative perspective: From IWRM to a 'nexus' approach? *Water Alternatives* **8** (1): 756–773

Berawi MA, Zagloel TY, Miraj P, Mulyanto H. 2017. Producing alternative concept for the Trans-Sumatera toll road project development using location quotient method. *Procedia Engineering* **171**: 265–273 DOI: 10.1016/j.proeng.2017.01.334

Biggs EM, Bruce E, Boruff B, Duncan JMA, Horsley J, Pauli N, McNeill K, Neef A, Van Ogtrop F, Curnow J, et al. 2015. Sustainable development and the water–energy–food nexus: A perspective on livelihoods. *Environmental Science & Policy* **54**: 389–397 DOI: 10.1016/J.ENVSCI.2015.08.002

Billings SB, Johnson EB. 2012. The location quotient as an estimator of industrial concentration. *Regional Science and Urban Economics* **42** (4): 642–647 DOI: 10.1016/j.regsciurbeco.2012.03.003

Binder T, Belyazid S, Haraldsson H V, Svensson MG, Kennedy M, Winch GW, Langer RS, Rowe JI, Yanni JM. 2004. Developing System Dynamics Models from Causal Loop Diagrams. In *Proceedings of the 22nd International Conference of the System Dynamics Society*Oxford.

Bizikova L, Roy D, Swanson D, Venema HD, McCandless M. 2013. The water–energy–food security nexus: Towards a practical planning and decision-support framework for landscape investment and risk management. International Institute for Sustainable Development (IISD), Manitoba.

BKP-RI. 2018. Indeks ketahanan pangan Indonesia 2018. Jakarta, Indonesia.

Blakely EJ, Bradshaw TK. 2002. *Planning local economic development: theory and practice*. SAGE Publications: California.

BPS of Bekasi. 2016. Bekasi City in Figures 2016. Bekasi, Indonesia.

BPS of Cianjur. 2016. Cianjur Regency in Figures 2016. Cianjur, Indonesia.

BPS of Karawang. 2015. Karawang Regency in Figures 2015. Karawang, Indonesia.

BPS of Karawang. 2016a. Karawang Regency in Figures 2016. Karawang, Indonesia.

BPS of Karawang. 2016b. Welfare Statistics of Karawang Regency (Statistik Kesejahteraan). Karawang, Indonesia.

BPS of Karawang. 2018. Karawang Regency in Figures 2018. Karawang. Indonesia.

BPS of Karawang. 2019. Karawang Regency in Figures 2019. Karawang, Indonesia.

BPS of West Java. 2016a. West Java Province in Figures 2016. Bandung, Indonesia.

BPS of West Java. 2016b. Gross Regional Domestic Product of West Java Province by Industrial Origin 2011-2015. Statistics of West Java Province, Bandung, Indonesia.

Brantingham PL, Brantingham PL. 1998. Mapping crime for analytic purposes: location quotients, counts and rates. *Crime mapping and crime prevention* (8): 263–288

Bréthaut C, Gallagher L, Dalton J, Allouche J. 2019. Power dynamics and integration in the water-energy-food nexus: Learning lessons for transdisciplinary research in Cambodia. *Environmental Science and Policy* **94** (January): 153–162 DOI: 10.1016/j.envsci.2019.01.010

Brodsky H, Sarfaty DE. 1977. Measuring the urban economic base in a developing country. *Land Economics* **53** (4): 445–454 Available at: http://www.jstor.org/stable/3145988

Brouwer F, Anzaldi G, Laspidou C, Munaretto S, Schmidt G, Strosser P, Sušnik J, Vamvakeridou-Lyroudia L. 2018. Commentary to SEI report 'Where is the added value? A review of the water-energy-food nexus literature' Available at: https://www.sim4nexus.eu/page.php?wert=Publications#collapse163

Cairns R, Krzywoszynska A. 2016. Anatomy of a buzzword: The emergence of 'the water-energy-food nexus' in UK natural resource debates. *Environmental Science and Policy* **64**: 164–170 DOI: 10.1016/j.envsci.2016.07.007

van Camp M, Mjemah IC, Al Farrah N, Walraevens K. 2013. Modeling approaches and strategies for data-scarce aquifers: example of the Dar es Salaam aquifer in Tanzania. *Hydrogeology Journal* **21** (2): 341–356 DOI: 10.1007/s10040-012-0908-5

Cansino-loeza B, Ponce-ortega M. 2020. Sustainable assessment of Water-Energy-Food Nexus at regional level through a multi-stakeholder optimization approach. (xxxx) DOI: 10.1016/j.jclepro.2020.125194

Chalar G, Garcia-Pesenti P, Silva-Pablo M, Perdomo C, Olivero V, Arocena R. 2017. Weighting the impacts to stream water quality in small basins devoted to forage crops, dairy and beef cow production. *Limnologica* **65** (January): 76–84 DOI: 10.1016/j.limno.2017.06.002

Chang SC. 2014. Effects of financial developments and income on energy consumption. *International Review of Economics and Finance* **35**: 28–44 DOI: 10.1016/j.iref.2014.08.011

Chiang SH. 2009. Location quotient and trade. *Annals of Regional Science* **43** (2): 399–414 DOI: 10.1007/s00168-008-0218-y

Daher B, Hannibal B, Mohtar RH, Portney K. 2020. Toward understanding the convergence of researcher and stakeholder perspectives related to water-energy-food (WEF) challenges: The case of San Antonio, Texas. *Environmental Science and Policy* **104** (April 2019): 20–35 DOI: 10.1016/j.envsci.2019.10.020

Daher BT, Mohtar RH. 2015. Water–energy–food (WEF) Nexus Tool 2.0: guiding integrative resource planning and decision-making. *Water International* (September 2015): 1–24 DOI: 10.1080/02508060.2015.1074148

Dai J, Wu S, Han G, Weinberg J, Xie X, Wu X, Song X, Jia B, Xue W, Yang Q. 2018. Water-energy nexus: A review of methods and tools for macro-assessment. *Applied Energy* **210** (September 2017): 393–408 DOI: 10.1016/j.apenergy.2017.08.243

Dang VL, Yeo GT. 2017. A Competitive strategic position analysis of major container ports in Southeast Asia. *Asian Journal of Shipping and Logistics* **33** (1): 1–10 DOI: 10.1016/j.ajsl.2017.03.003

Dargin J, Daher B, Mohtar RH. 2019. Complexity versus simplicity in water energy food nexus (WEF) assessment tools. *Science of the Total Environment* **650** (2019): 1566–1575 DOI: 10.1016/j.scitotenv.2018.09.080

Das K, Gerbens-leenes PW, Nonhebel S. 2020. The water footprint of food and cooking fuel : A case study of self- suf fi cient rural India. *Journal of Cleaner Production* (xxxx): 125255 DOI: 10.1016/j.jclepro.2020.125255

Davies EGR, Kyle P, Edmonds JA. 2013. An integrated assessment of global and regional water demands for electricity generation to 2095. *Advances in Water Resources* **52**: 296–313 DOI: 10.1016/j.advwatres.2012.11.020

Day J, Ellis P. 2012. Growth in Indonesia's manufacturing sectors: urban and localization contributions. In *Sixth Urban Research and Knowledge Symposium*.

Dinar A, Tieu A, Huynh H. 2019. Water scarcity impacts on global food production. *Global Food Security* **23** (July): 212–226 DOI: 10.1016/j.gfs.2019.07.007

Dodder RS. 2014. A review of water use in the U.S. electric power sector: Insights from systems-level perspectives. *Current Opinion in Chemical Engineering* **5**: 7–14 DOI: 10.1016/j.coche.2014.03.004

Elosegi A, Sabater S. 2013. Effects of hydromorphological impacts on river ecosystem functioning: A review and suggestions for assessing ecological impacts. *Hydrobiologia* **712** (1): 129–143 DOI: 10.1007/s10750-012-1226-6

Elsoragaby S, Yahya A, Mahadi MR, Nawi NM, Mairghany M. 2019. Energy utilization in major crop cultivation. *Energy* **173**: 1285–1303 DOI: 10.1016/j.energy.2019.01.142

Emas R. 2015. The concept of sustainable development: definition and defining principles. *Brief for GSDR*: 1–3

Endo A, Burnett K, Orencio PM, Kumazawa T, Wada CA, Ishii A, Tsurita I, Taniguchi M. 2015. Methods of the water-energy-food nexus. *Water (Switzerland)* **7** (10): 5806–5830 DOI: 10.3390/w7105806

Endo A, Tsurita I, Burnett K, Orencio PM. 2017. A review of the current state of research on the water, energy, and food nexus. *Journal of Hydrology: Regional Studies* **11**: 20–30 DOI: 10.1016/j.ejrh.2015.11.010

Endo A, Yamada M, Miyashita Y, Sugimoto R, Ishii A, Nishijima J, Fujii M, Kato T, Hamamoto H, Kimura M, et al. 2019. Dynamics of water–energy–food nexus methodology, methods, and tools. *Current Opinion in Environmental Science and Health* **13**: 46–60 DOI: 10.1016/j.coesh.2019.10.004

European Report on Development. 2012. Confronting scarcity: Managing water, energy and land for inclusive and sustainable growth. Belgium. DOI: 10.2841/40899

Falkenmark M, Lundqvist J, Widstrand C. 1989. Macro-scale water scarcity requires micro-scale approaches: Aspects of vulnerability in semi-arid development. *Natural Resources Forum* **13** (4): 258–267 DOI: 10.1111/j.1477-8947.1989.tb00348.x

Fan H, He D, Wang H. 2015. Environmental consequences of damming the mainstream lancang-mekong river: A review. *Earth-Science Reviews* **146** (2): 77–91 DOI: 10.1016/j.earscirev.2015.03.007

FAO. 2014. The water-energy-food nexus - A new approach in support of food security and sustainable agriculture. *Food and Agriculture Organization of the United Na*: 1–11 DOI: 10.1039/C4EW90001D

FAO. 2018. Accelerating SDG 7 achievement. Policy Brief 09. Water-Energy-Food Nexus for the review of SDG 7 Available at: https://sustainabledevelopment.un.org/content/documents/17480PB8.pdf

Feng K, Hubacek K, Siu YL, Li X. 2014. The energy and water nexus in Chinese electricity production: A hybrid life cycle analysis. *Renewable and Sustainable Energy Reviews* **39**: 342–355 DOI: 10.1016/j.rser.2014.07.080

Feng M, Liu P, Li Z, Zhang J, Liu D, Xiong L. 2016. Modeling the nexus across water supply, power generation and environment systems using the system dynamics approach: Hehuang Region, China. *Journal of Hydrology* DOI: 10.1016/j.jhydrol.2016.10.011

Flammini A, Puri M, Pluschke L, Dubois O. 2014. *Walking the nexus talk: Assessing the water-energy-food nexus in the context of the sustainable energy for all initiative*. Food and Agriculture Organization (FAO): Rome. Available at: http://www.fao.org/publications/card/en/c/f065f1d5-2dda-4df7-8df3-4defb5a098c8/

Foran T. 2015. Node and regime: Interdisciplinary analysis of water-energy-food nexus in the Mekong region. *Water Alternatives* **8** (1): 655–674

Forrester JW. 1961. *Industrial Dynamics*. MIT Press, Cambridge: Massachushetts.

de Fraiture C, Wichelns D. 2010. Satisfying future water demands for agriculture. *Agricultural Water Management* **97** (4): 502–511 DOI: 10.1016/j.agwat.2009.08.008

de Fraiture C, Fayrap A, Unver O, Ragab R. 2014. Integrated water management approaches for sustainable food production. *Irrigation and Drainage* **63** (2): 221–231 DOI: 10.1002/ird.1847

de Fraiture C, Molden D, Wichelns D. 2010. Investing in water for food, ecosystems, and livelihoods: An overview of the comprehensive assessment of water management in agriculture. *Agricultural Water Management* **97** (4): 495–501 DOI: 10.1016/j.agwat.2009.08.015

El Gafy I, Grigg N, Reagan W. 2016. Dynamic behaviour of the water-food-energy nexus: Focus on crop production and consumption. *Irrigation and Drainage* DOI: 10.1002/ird.2060

Galaitsi S, Veysey J, Huber-lee A. 2018. Where is the added value? A review of the water-energy-food nexus literature

Gamage IU, Jayasena HAH. 2018. Socio-hydrological implications of water management in the dry zone of Sri Lanka. *Proceedings of the International Association of Hydrological Sciences* **379**: 415–420 DOI: 10.5194/piahs-379-415-2018

van Gevelt T. 2020. The water–energy–food nexus: bridging the science–policy divide. *Current Opinion in Environmental Science & Health* **13**: 6–10 DOI: 10.1016/J.COESH.2019.09.008

Ghashghaie M, Marofi S, Marofi H. 2014. Using System Dynamics Method to Determine the Effect of Water Demand Priorities on Downstream Flow DOI: 10.1007/s11269-014-0791-z

Goetz S, Deller S, Harris T. 2007. Targeting regional economic development: an outline of a national extension educational program. In *The 2007 CDS Annual Meetings* The Northeast Regional Center for Rural Development: Pennsylvania.

Goh YM, Love PED, Stagbouer G, Annesley C. 2012. Dynamics of safety performance and culture: A group model building approach. *Accident Analysis and Prevention* **48**: 118–125 DOI: 10.1016/j.aap.2011.05.010

de Grenade R, House-Peters L, Scott CA, Thapa B, Mills-Novoa M, Gerlak A, Verbist K. 2016. The nexus: reconsidering environmental security and adaptive capacity. *Current Opinion in Environmental Sustainability* **21**: 15–21 DOI: 10.1016/ j.cosust.2016.10.009

Guo HC, Liu L, Huang GH, Fuller GA, Zou R, Yin YY. 2001. A system dynamics approach for regional environmental planning and management: a study for the Lake Erhai Basin. *Journal of environmental management* **61** (1): 93–111 DOI: 10.1006/jema.2000.0400

Haraldson H V. 2000. *Introduction to Systems and Causal Loop Diagrams.* Lund University: Lund, Sweden. Available at: http://dev.crs.org.pl:4444/ rid=1244140954250_1167059429_1461/Introduction to Systems and Causal Loop Diagrams.pdf

Hecht JS, Lacombe G, Arias ME, Dang TD, Piman T. 2019. Hydropower dams of the Mekong River basin: A review of their hydrological impacts. *Journal of Hydrology* **568** (October 2018): 285–300 DOI: 10.1016/j.jhydrol.2018.10.045

Hoekstra AY. 2017. Water Footprint Assessment: Evolvement of a New Research Field. *Water Resources Management* **31** (10): 3061–3081 DOI: 10.1007/s11269-017- 1618-5

Hoff H. 2011. Understanding the nexus. In *Background Paper for the Bonn2011 Conference: The Water, Energy and Food Security Nexus.* Stockholm Environment Institute: Stockholm.

Hoff H, Alrahaife SA, El Hajj R, Lohr K, Mengoub FE, Farajalla N, Fritzsche K, Jobbins G, özerol G, Schultz R, et al. 2019. A nexus approach for the MENA region-from concept to knowledge to action. *Frontiers in Environmental Science* **7** (APR): 1– 14 DOI: 10.3389/fenvs.2019.00048

Hooda PS, Edwards AC, Anderson HA, Miller A. 2000. A review of water quality concerns in livestock farming areas. *Science of the Total Environment* **250** (1–3): 143–167 DOI: 10.1016/S0048-9697(00)00373-9

Horta IM, Camanho AS. 2014. Competitive positioning and performance assessment in the construction industry. *Expert Systems with Applications* **41** (4 PART 1): 974– 983 DOI: 10.1016/j.eswa.2013.06.064

Hovmand PS, Andersen DF, Rouwette E, Richardson GP, Rux K, Calhoun A. 2012. Group Model-Building 'Scripts' as a Collaborative Planning Tool. *Systems Research and Behavioral Science* **29**: 179–193 DOI: 10.1002/sres

Howells M, Hermann S, Welsch M, Bazilian M, Segerström R, Alfstad T, Gielen D, Rogner H, Fischer G, Van Velthuizen H, et al. 2013. Integrated analysis of climate change, land-use, energy and water strategies. *Nature Climate Change* **3** (7): 621– 626 DOI: 10.1038/nclimate1789

Hülsmann S, Sušnik J, Rinke K, Langan S, van Wijk D, Janssen AB, Mooij WM. 2019. Integrated modelling and management of water resources: the ecosystem perspective on the nexus approach. *Current Opinion in Environmental Sustainability* **40**: 14–20 DOI: 10.1016/j.cosust.2019.07.003

ICIMOD. 2012. Contribution of Himalayan ecosystems to water, energy and food security in South Asia: a nexus approach. Kathmandu, Nepal.

IESR. 2019. Powering the cities: Perhitungan potensi PLTS atap untuk bangunan Pemerintah di Sumatera Utara, DKI Jakarta, Jawa Tengah, Surabaya, dan Bali

IFPRI. 2016. Global Hunger Index 2016: Getting to zero hunger. International Food Policy Research Institute, Washington DC. DOI: http://dx.doi.org/10.2499/9780896292260

Inam A, Adamowski J, Halbe J, Prasher S. 2015. Using causal loop diagrams for the initialization of stakeholder engagement in soil salinity management in agricultural watersheds in developing countries: A case study in the Rechna Doab watershed, Pakistan. *Journal of Environmental Management* **152**: 251–267 DOI: 10.1016/j.jenvman.2015.01.052

IRENA. 2015. Renewable energy in the water, energy and food nexus. International Renewable Energy Agency, Abu Dhabi.

Islam F Bin, Siddiq F, Mubassirah FA. 2016. Economic growth analysis of six divisions of Bangladesh using location quotient and shift-share method. *Bangladesh Research Publication Journal* **12** (December): 144–154

Isserman AM. 1977. The location quotient approach to estimating regional economic impacts. *Journal of the American Planning Association* **43** (1): 33–41 DOI: 10.1080/01944367708977758

Karnib A. 2018. Bridging Science and Policy in Water-Energy-Food Nexus: Using the Q-Nexus Model for Informing Policy Making. *Water Resources Management* **32** (15): 4895–4909 DOI: 10.1007/s11269-018-2059-5

Keairns DL, Darton RC, Irabien A. 2016. The Energy-Water-Food Nexus. *Annual Review of Chemical and Biomolecular Engineering* **7**: 239–262 DOI: 10.1146/annurev-chembioeng-080615-033539

Kenway SJ, Lant PA, Priestley A, Daniels P. 2011. The connection between water and energy in cities: A review. *Water Science and Technology* **63** (9) DOI: 10.2166/wst.2011.070

Kotir JH, Brown G, Marshall N, Johnstone R. 2017. Environmental Modelling & Software Systemic feedback modelling for sustainable water resources management and agricultural development : An application of participatory modelling approach in the Volta River Basin. *Environmental Modelling and Software* **88**: 106–118 DOI: 10.1016/j.envsoft.2016.11.015

Kotir JH, Smith C, Brown G, Marshall N, Johnstone R. 2016. A system dynamics simulation model for sustainable water resources management and agricultural development in the Volta River Basin, Ghana. *Science of the Total Environment* **573**: 444–457 DOI: 10.1016/j.scitotenv.2016.08.081

Kummu M, Guillaume JHA, De Moel H, Eisner S, Flörke M, Porkka M, Siebert S, Veldkamp TIE, Ward PJ. 2016. The world's road to water scarcity: Shortage and stress in the 20th century and pathways towards sustainability. *Scientific Reports* **6** (December) DOI: 10.1038/srep38495

Lawford RG. 2019. A design for a data and information service to address the knowledge needs of the Water-Energy-Food (W-E-F) Nexus and strategies to facilitate its implementation. *Frontiers in Environmental Science* **7** (APR) DOI: 10.3389/fenvs.2019.00056

Leck H, Conway D, Bradshaw M, Rees J. 2015. Tracing the water energy food nexus: Description, theory and practice. *Geography Compass (in press)* **8**: 445–460 DOI: 10.1111/gec3.12222

Leese M, Meisch S. 2015. Securitising sustainability? Questioning the 'water, energy and food-security nexus'. *Water Alternatives* **8** (1): 695–709

Leigh R. 1970. The use of location quotients in urban economic base studies. *Land Economics* **46** (2): 202–205 DOI: 10.2307/3145181

Lele U, Klousia-Marquis M, Goswami S. 2013. Good Governance for Food, Water and Energy Security. *Aquatic Procedia* **1**: 44–63 DOI: 10.1016/j.aqpro.2013.07.005

Lezzaik K, Milewski A. 2018. A quantitative assessment of groundwater resources in the Middle East and North Africa region. *Hydrogeology Journal* **26** (1): 251–266 DOI: 10.1007/s10040-017-1646-5

Liang Y, Cai W, Ma M. 2019. Carbon dioxide intensity and income level in the Chinese megacities' residential building sector: Decomposition and decoupling analyses. *Science of The Total Environment* **677**: 315–327 DOI: 10.1016/J.SCITOTENV.2019.04.289

Liao X, Hall JW, Eyre N. 2016. Water use in China's thermoelectric power sector. *Global Environmental Change* **41**: 142–152 DOI: 10.1016/j.gloenvcha.2016.09.007

Liermann CR, Nilsson C, Robertson J, Ng RY. 2012. Implications of dam obstruction for global freshwater fish diversity. *BioScience* **62** (6): 539–548 DOI: 10.1525/bio.2012.62.6.5

Liu J, Yang H, Cudennec C, Gain AK, Hoff H, Lawford R, Qi J, de Strasser L, Yillia PT, Zheng C. 2017. Challenges in operationalizing the water–energy–food nexus. *Hydrological Sciences Journal* **62** (11) DOI: 10.1080/02626667.2017.1353695

Liu Y, Chen B. 2020. Water-energy scarcity nexus risk in the national trade system based on multiregional input-output and network environ analyses. *Applied Energy* **268** (19): 114974 DOI: 10.1016/j.apenergy.2020.114974

Livingstone I. 1968. Agriculture versus industry in economic development agriculture versus industry in economic development. *The Journal of Modern African Studies* **6** (3): 329–341

Love BJ, Nejadhashemi AP. 2011. Water quality impact assessment of large-scale biofuel crops expansion in agricultural regions of Michigan. *Biomass and Bioenergy* **35** (5): 2200–2216 DOI: 10.1016/j.biombioe.2011.02.041

Luna-Reyes LF, Martinez-Moyano IJ, Pardo TA, Cresswell AM, Andersen DF, Richardson GP. 2006. Anatomy of a Group Model-Building Intervention: Building Dynamic Theory from case Study research. *Wiley InterScience* **22** (4): 291–320 DOI: 10.1002/sdr.349

Ma M, Cai W, Cai W, Dong L. 2019. Whether carbon intensity in the commercial building sector decouples from economic development in the service industry? Empirical evidence from the top five urban agglomerations in China. *Journal of Cleaner Production* **222**: 193–205 DOI: 10.1016/J.JCLEPRO.2019.01.314

MacDonald AM, Bonsor HC, Dochartaigh BÉÓ, Taylor RG. 2012. Quantitative maps of groundwater resources in Africa. *Environmental Research Letters* **7** (2) DOI: 10.1088/1748-9326/7/2/024009

Macknick J, Newmark R, Heath G, Hallett KC. 2012. Operational water consumption and withdrawal factors for electricity generating technologies: A review of existing literature. *Environmental Research Letters* **7** (4) DOI: 10.1088/1748-9326/7/4/045802

Mannan M, Al-Ansari T, Mackey HR, Al-Ghamdi SG. 2018. Quantifying the energy, water and food nexus: A review of the latest developments based on life-cycle assessment. *Journal of Cleaner Production* **193**: 300–314 DOI: 10.1016/j.jclepro.2018.05.050

Märker C, Venghaus S, Hake JF. 2018. Integrated governance for the food–energy–water nexus – The scope of action for institutional change. *Renewable and Sustainable Energy Reviews* **97** (September): 290–300 DOI: 10.1016/j.rser.2018.08.020

Martinez P, Blanco M, Castro-Campos B. 2018. The Water–Energy–Food Nexus: A Fuzzy-Cognitive Mapping Approach to Support Nexus-Compliant Policies in Andalusia (Spain). *Water* **10** (5): 664 DOI: 10.3390/w10050664

Masella P, Galasso I. 2020. A comparative cradle-to-gate life cycle study of bio-energy feedstock from camelina Sativa, an Italian case study. *Sustainability (Switzerland)* **12** (22): 1–21 DOI: 10.3390/su12229590

McCarl BA, Yang Y, Srinivasan R, Pistikopoulos EN, Mohtar RH. 2017. Data for WEF Nexus Analysis: a Review of Issues. *Current Sustainable/Renewable Energy Reports* **4** (3): 137–143 DOI: 10.1007/s40518-017-0083-3

Meadow DH. 2009. *Thinking in System: A Primer* (D Wright, ed.). Chelsea Green Publishing: Vermont.

Mekonnen MM, Gerbens-Leenes PW, Hoekstra AY. 2015. The consumptive water footprint of electricity and heat: A global assessment. *Environmental Science: Water Research and Technology* **1** (3): 285–297 DOI: 10.1039/c5ew00026b

Mercure JF, Paim MA, Bocquillon P, Lindner S, Salas P, Martinelli P, Berchin II, de Andrade Guerra JBSO, Derani C, de Albuquerque Junior CL, et al. 2019. System complexity and policy integration challenges: The Brazilian Energy- Water-Food Nexus. *Renewable and Sustainable Energy Reviews* **105** (March 2018): 230–243 DOI: 10.1016/j.rser.2019.01.045

Mguni P, van Vliet B, Spaargaren G, Nakirya D, Osuret J, Isunju JB, Ssekamatte T, Mugambe R. 2020. What could go wrong with cooking? Exploring vulnerability at the water, energy and food Nexus in Kampala through a social practices lens. *Global Environmental Change* **63** (April): 102086 DOI: 10.1016/j.gloenvcha.2020.102086

Middleton C, Allouche J, Gyawali D, Allen S. 2015. The rise and implications of the water-energy-food nexus in Southeast Asia through an environmental justice lens. *Water Alternatives* **8** (1)

Miller MM, Gibson LJ, Wright NG. 1991. Location quotient: a basic tool for economic development analysis. *Economic development review* **9** (2): 65–68 Available at: http://search.proquest.com/openview/d6011b83d027b7ad1dba29bb96b74a53/1.pdf?pq-origsite=gscholar&cbl=38209

Mirchi A, Madani K, Watkins D, Ahmad S. 2012. Synthesis of System Dynamics Tools for Holistic Conceptualization of Water Resources Problems. *Water Resources Management* **26** (9): 2421–2442 DOI: 10.1007/s11269-012-0024-2

Mitchell B, Bellette K, Richardson S. 2015. 'Integrated' approaches to water and natural resources management in South Australia. *International Journal of Water Resources Development* **31** (4): 718–731 DOI: 10.1080/07900627.2014.979399

MoE-RI. 2018. Indeks Kualitas Lingkungan Hidup Indonesia 2017. Jakarta, Indonesia. DOI: 10.1093/nar/4.6.1727

Moghaddasi R, Pour AA. 2016. Energy consumption and total factor productivity growth in Iranian agriculture. *Energy Reports* **2**: 218–220 DOI: 10.1016/j.egyr.2016.08.004

Morrissey K. 2014. Producing regional production multipliers for Irish marine sector policy: A location quotient approach. *Ocean & Coastal Management* **91** (3): 58–64 DOI: 10.1111/pirs.12143

Muhammad A, D'Souza A, Meade B, Micha R, Mozaffarian D. 2017. How income and food prices influence global dietary intakes by age and sex: Evidence from 164 countries. *BMJ Global Health* **2** (3): 1–11 DOI: 10.1136/bmjgh-2016-000184

Mushayagwazvo MT. 2020. Spatial water, energy, and food (WEF) nexus exploration in a local context: A case of Karawang Regency, Indonesia.IHE Delft, Institute for Water Education.

Nadia Putri Utami, Ahamed T. 2018. Spatial Analysis to Determine Paddy Field Changes in Indonesia : A Case Study in Suburban Areas of Jakarta. In *Earth and Environmental Science*0–7.

Nhamo L, Mabhaudhi T, Mpandeli S, Dickens C, Nhemachena C, Senzanje A, Naidoo D, Liphadzi S, Modi AT. 2020. An integrative analytical model for the water-energy-food nexus: South Africa case study. *Environmental Science and Policy* **109** (April): 15–24 DOI: 10.1016/j.envsci.2020.04.010

Nouri H, Stokvis B, Chavoshi Borujeni S, Galindo A, Brugnach M, Blatchford ML, Alaghmand S, Hoekstra AY. 2020. Reduce blue water scarcity and increase nutritional and economic water productivity through changing the cropping pattern in a catchment. *Journal of Hydrology* **588** (June): 125086 DOI: 10.1016/j.jhydrol.2020.125086

Nyathi M, Annandale J, Beletse Y, Beukes D, Plooy C du, Pretorius B, Halsema G van. 2016. Nutritional water productivity of traditional vegetable crops. Gezina.

Nyathi MK, Van Halsema GE, Beletse YG, Annandale JG, Struik PC. 2018. Nutritional water productivity of selected leafy vegetables. *Agricultural Water Management* **209** (December 2017): 111–122 DOI: 10.1016/j.agwat.2018.07.025

Nyathi MK, Mabhaudhi T, Van Halsema GE, Annandale JG, Struik PC. 2019. Benchmarking nutritional water productivity of twenty vegetables - A review. *Agricultural Water Management* **221** (January): 248–259 DOI: 10.1016/j.agwat.2019.05.008

Odiyo JO, Chimuka L, Mamali MA, Fatoki OS. 2012. Trophic status of Vondo and Albasini Dams; impacts on aquatic ecosystems and drinking water. *International Journal of Environmental Science and Technology* **9** (2): 203–218 DOI: 10.1007/s13762-012-0034-x

OECD. 2017. The land-water-energy nexus: Biophysical and economic consequences. OECD Publishing, Paris. DOI: 10.2166/9781780409283

Okadera T, Geng Y, Fujita T, Dong H, Liu Z, Yoshida N, Kanazawa T. 2015. Evaluating the water footprint of the energy supply of Liaoning Province, China: A regional input-output analysis approach. *Energy Policy* **78**: 148–157 DOI: 10.1016/j.enpol.2014.12.029

Pacetti T, Lombardi L, Federici G. 2015. Water-energy Nexus: A case of biogas production from energy crops evaluated by Water Footprint and Life Cycle Assessment (LCA) methods. *Journal of Cleaner Production* **101**: 278–291 DOI: 10.1016/j.jclepro.2015.03.084

Pahl-Wostl C. 2019. Governance of the water-energy-food security nexus: A multi-level coordination challenge. *Environmental Science and Policy* **92** (January 2017): 356–367 DOI: 10.1016/j.envsci.2017.07.017

Pan L, Liu P, Li Z. 2017. A system dynamic analysis of China's oil supply chain: Overcapacity and energy security issues. *Applied Energy* **188**: 508–520 DOI: 10.1016/j.apenergy.2016.12.036

Pan S-Y, Snyder SW, Packman AI, Lin YJ, Chiang P-C. 2018. Cooling water use in thermoelectric power generation and its associated challenges for addressing water-energy nexus. *Water-Energy Nexus* **1** (1): 26–41 DOI: 10.1016/j.wen.2018.04.002

Pearce JA, Robinson RB. 2001. *Strategic management: formulation, implementation, and control*. Mc Graw Hill: New York.

Prasad G, Stone A, Hughes A, Stewart T. 2012. Towards the Development of an Energy-Water-Food Security Nexus based Modelling Framework as a Policy and Planning Tool for South Africa. *Strategies to Overcome Poverty & Inequality* Available at: http://www.carnegie3.org.za/docs/papers/255_Prasad_Towards the development of an energy-water-food security nexus based modelling framework as a policy and planning tool for SA.pdf

Purwanto A, Sušnik J, Suryadi FX, de Fraiture C. 2018. Determining strategies for water, energy, and food-related sectors in local economic development. *Sustainable Production and Consumption* **16** DOI: 10.1016/j.spc.2018.08.005

Purwanto A, Sušnik J, Suryadi FX, de Fraiture C. 2019. Using group model building to develop a causal loop mapping of the water-energy-food security nexus in Karawang Regency, Indonesia. *Journal of Cleaner Production* DOI: 10.1016/j.jclepro.2019.118170

Purwanto A, Sušnik J, Suryadi FX, de Fraiture C. 2020a. Quantitative simulation of the water-energy-food (WEF) security nexus in a local planning context in indonesia. *Sustainable Production and Consumption* **25**: 198–216 DOI: 10.1016/j.spc.2020.08.009

Purwanto A, Sušnik J, Suryadi FX, De Fraiture C. 2020b. Supplementary materials. *Sustainable Production and Consumption* DOI: http://doi:10.1016/j.spc.2020.08.009

Quincieu E. 2015. Summary of Indonesia's agriculture, natural resources, and environment sector assessment. Manila, Philippines.

Rafiuddin A, Widiatmaka W, Munibah K. 2016. Pola perubahan penggunaan lahan dan neraca pangan di Kabupaten Karawang. *Journal Ilmu Pertanian dan Lingkungan* **I** (April): 15–20

Rahmadi A. 2013. Indonesia at crossroad: Addressing food security Available at: www.arahmadi.net/energi/rahmadi-addressing-food-security-2013.pdf [Accessed 4 December 2016]

Rahmati O, Pourghasemi HR, Melesse AM. 2016. Application of GIS-based data driven random forest and maximum entropy models for groundwater potential mapping: A case study at Mehran Region, Iran. *Catena* **137**: 360–372 DOI: 10.1016/j.catena.2015.10.010

Ramos E, Kofinas D, Chrysaida Papadopoulou, Papadopoulou M, Gardumi F, Brouwer F, Fournier M, Echeverria L, Domingo X, Vamvakeridou-Lyroudia L, et al. 2020. D1.5 : Framework for the assessment of the nexus

Rastegari S, Ehsan M, Piewthongngam K. 2000. An analysis of industrial – agricultural interactions : a case study in Pakistan. **22**: 17–27

Rautaray SK, Pradhan S, Mohanty S, Dubey R, Raychaudhuri S, Mohanty RK, Mishra A, Ambast SK. 2020. Energy efficiency, productivity and profitability of rice farming using Sesbania as green manure-cum-cover crop. *Nutrient Cycling in Agroecosystems* **116** (1): 83–101 DOI: 10.1007/s10705-019-10034-z

Reinhard S, Verhagen J, Wolters W, Ruben R. 2017. Water-food-energy nexus: 28 Available at: https://library.wur.nl/WebQuery/wurpubs/fulltext/424551

Rezaei A, Mohammadi Z. 2017. Annual safe groundwater yield in a semiarid basin using combination of water balance equation and water table fluctuation. *Journal of African Earth Sciences* **134**: 241–248 DOI: 10.1016/j.jafrearsci.2017.06.029

Rich KM, Rich M, Dizyee K. 2018. Participatory systems approaches for urban and peri-urban agriculture planning: The role of system dynamics and spatial group model building. *Agricultural Systems* **160**: 110–123 DOI: 10.1016/j.agsy.2016.09.022

Richardson GP, Andersen DF. 2010. Systems Thinking, Mapping, and Modeling in Group Decision and Negotiation: 1–19

Richardson HW. 1985. Input-output and economic base multipliers: looking backward and forward. *Journal of Regional Science* **25** (4): 607–661 DOI: 10.1111/j.1467-9787.1985.tb00325.x

Riddington G, Gibson H, Anderson J. 2006. Comparison of gravity model, survey and location quotient-based local area tables and multipliers. *Regional Studies* **40** (9): 1069–1081 DOI: 10.1080/00343400601047374

Ringler C, Bhaduri A, Lawford R. 2013. The nexus across water, energy, land and food (WELF): Potential for improved resource use efficiency? *Current Opinion in Environmental Sustainability* **5** (6): 617–624 DOI: 10.1016/j.cosust.2013.11.002

Sadegh M, AghaKouchak A, Mallakpour I, Huning LS, Mazdiyasni O, Niknejad M, Foufoula-Georgiou E, Moore FC, Brouwer J, Farid A, et al. 2020. Data and analysis toolbox for modeling the nexus of food, energy, and water. *Sustainable Cities and Society* **61** (April) DOI: 10.1016/j.scs.2020.102281

Shannak S, Mabrey D, Vittorio M. 2018. Moving from theory to practice in the water–energy–food nexus: An evaluation of existing models and frameworks. *Water-Energy Nexus* **1** (1): 17–25 DOI: 10.1016/j.wen.2018.04.001

Simmonds K. 1986. The accounting assessment of competitive position. *European Journal of Marketing* **20** (1): 16–31 DOI: 10.1108/EUM0000000004626

Simonovic SP. 2009. *Managing Water Resources: Methods and Tools for A System Approach*. UNESCO Publishing & Earthscan: London.

Simpson GB, Jewitt GP. 2019. The water-energy-food nexus in the anthropocene: moving from 'nexus thinking' to 'nexus action'. *Current Opinion in Environmental Sustainability* **40**: 117–123 DOI: 10.1016/j.cosust.2019.10.007

Smajgl A, Ward J, Pluschke L. 2016. The water–food–energy Nexus – Realising a new paradigm. *Journal of Hydrology* **533**: 533–540 DOI: 10.1016/j.jhydrol.2015.12.033

Sperling JB, Berke PR. 2017. Urban nexus science for future cities: Focus on the energy-water-food-X nexus. Texas.

Sterman JD. 2000. *Business Dynamics: Systems Thinking and Modeling for a Complex World*. Mc Graw Hill India: New Delhi. DOI: 10.1108/13673270210417646

Sun Y, Liu N, Shang J, Zhang J. 2015. Sustainable utilization of water resources in China: A system dynamics model. *Journal of Cleaner Production* **142**: 613–625 DOI: 10.1016/j.jclepro.2016.07.110

Sušnik J. 2018. Data-driven quantification of the global water-energy-food system. *Resources, Conservation and Recycling* **133**: 179–190 DOI: 10.1016/J.RESCONREC.2018.02.023

Sušnik J, Chew C, Domingo X, Mereu S, Trabucco A, Evans B, Vamvakeridou-Lyroudia L, Savić DA, Laspidou C, Brouwer F. 2018. Multi-stakeholder development of a serious game to explore the water-energy-food-land-climate nexus: The SIM4NEXUS approach. *Water (Switzerland)* **10** (2) DOI: 10.3390/w10020139

Sušnik J, Vamvakeridou-Lyroudia LS, Savić DA, Kapelan Z. 2012. Integrated System Dynamics Modelling for water scarcity assessment: Case study of the Kairouan region. *Science of the Total Environment* **440**: 290–306 DOI: 10.1016/j.scitotenv.2012.05.085

Sušnik J, Vamvakeridou-Lyroudia LS, Savić DA, Kapelan Z. 2013. Integrated modelling of a coupled water-agricultural system using system dynamics. *Journal of Water and Climate Change* **4** (3): 209–231 DOI: 10.2166/wcc.2013.069

Sustainable Integrated Management FOR the NEXUS of water-landfood-energy-climate for a resource-efficient Europe (SIM4NEXUS). Available at: https://www.sim4nexus.eu/ [Accessed 4 December 2020]

Suyatno. 2000. Analisa economic base terhadap pertumbuhan ekonomi daerah tingkat II Wonogiri: menghadapi implementasi UU No. 22/1999 dan UU No. 5/1999 (in Bahasa Indonesia) DOI: 10.23917/jep.v1i2.3899

The Economist Intelligence Unit. 2016. Global food security index 2016: An annual measure of the state of global food security. The Economist Intelligence Unit Limited, London. Available at: http://foodsecurityindex.eiu.com/Country/Details#United%5CnStates

Tohmo T. 2004. New developments in the use of location quotients to estimate regional input – output coefficients and multipliers. *Regional Studies* (May 2013): 38:1, 43–54

Trappey AJC, Trappey CV, Liu PHY, Hsiao CT, J.J.R O, K.W.P. C. 2013. Location quotient EIO-LCA method for carbon emission analysis. In *Concurrent Engineering Approaches for Sustainable Product Development in a Multi-Disciplinary Environment*, Stjepandic J (ed.).Springer-Verlag London; 367–377. DOI: 10.1007/978-1-4471-4426-7

UN-Water. 2013. *Water security and the global water agenda: A UN-Water analytical brief.* United Nation University: Ontario. DOI: 10.1016/0022-1694(68)90080-2

UN-WWAP. 2015. The United Nations World Water Development Report 2015: Water for a sustainable world DOI: 978-92-3-100071-3

UNDP. 2016. Human Development Report 2016: Human development for everyone: 1–8

UNECE. 2015. Reconciling resource uses in transboundary basins: assessment of the water-food-energy-ecosystems nexus Available at: http://www.unece.org/fileadmin/DAM/env/water/publications/WAT_Nexus/ece_mp.wat_46_eng.pdf

United Nation. 1987. *Report of the world commission on environment and development: Our common future.* DOI: 10.1080/07488008808408783

Urbinatti AM, Benites-Lazaro LL, Carvalho CM de, Giatti LL. 2020. The conceptual basis of water-energy-food nexus governance: systematic literature review using network and discourse analysis. *Journal of Integrative Environmental Sciences* **17** (2): 21–43 DOI: 10.1080/1943815X.2020.1749086

Vanham D, Medarac H, Schyns JF, Hogeboom RJ, Magagna D. 2019. The consumptive water footprint of the European union energy sector. *Environmental Research Letters* **14** (10) DOI: 10.1088/1748-9326/ab374a

Vennix JAM. 1996. *Group Model Building: Facilitating Team Learning Using System Dynamics*. John Wiley & Sons Inc.

Vennix JAM, Akkermans HA, Rouwette EAJA. 1996. Group model-building to facilitate organizational change: an exploratory study. *System Dynamics Review* **12** (1): 39–58

Villamayor-Tomas S, Grundmann P, Epstein G, Evans T, Kimmich C. 2015. The water-energy-food security nexus through the lenses of the value chain and the institutional analysis and development frameworks. *Water Alternatives*

Walters WH. 2017. Citation-Based Journal Rankings: Key Questions, Metrics, and Data Sources. *IEEE Access* **5** (Section V): 22036–22053 DOI: 10.1109/ ACCESS.2017.2761400

Waluyo Hatmoko, Radhika, Asyantina T. 2017. Indeks ketahanan air pada wilayah sungai di Indonesia. Bandung, Indonesia.

Wang X, Hofe R. 2007. *Research methods in urban and regional planning*. Tsinghua University Press and Springer: Beijing.

Wang YB, Wu PT, Engel BA, Sun SK. 2014. Application of water footprint combined with a unified virtual crop pattern to evaluate crop water productivity in grain production in China. *Science of the Total Environment* **497–498**: 1–9 DOI: 10.1016/j.scitotenv.2014.07.089

WEF. 2011. Global Risks 2011 six edition: An initiative of the risk response network. Geneva, Switzerland. Available at: http://reports.weforum.org/wp-content/blogs.dir/1/mp/uploads/pages/files/global-risks-2011.pdf

Weitz N, Huber-Lee A, Nilsson M, Davis M, Hoff H. 2014. Cross-sectoral integration in the Sustainable Development Goals: A nexus approach: 8

WHO. 2017. Guidelines for drinking-water quality: fourth edition incorporating the first addendum. Geneva. Available at: https://www.who.int/publications/i/item/ 9789241549950

Wichelns D. 2017. The water-energy-food nexus: Is the increasing attention warranted, from either a research or policy perspective? *Environmental Science and Policy* **69**: 113–123 DOI: 10.1016/j.envsci.2016.12.018

Widiatmaka, Ambarwulan W, Munibah K. 2013. Landuse change during a decade as determined by landsat imagery of a rice production region and its implication to regional contribution to rice self sufficiency: Case study of karawang regency, west java, Indonesia. In *34th Asian Conference on Remote Sensing 2013, ACRS 2013*3330–3336. Available at: http://www.scopus.com/inward/record.url?eid=2-s2.0-84903433679&partnerID=40&md5=a77bd3030bfe80095418249d01ec45cc

Willis HH, Groves DG, Ringel JS, Mao Z, Efron S, Abbott M. 2016. Developing the Pardee RAND Food-Energy-Water Security Index: 60 DOI: 10.7249/TL165

Winz I, Brierley G, Trowsdale S. 2009. The use of system dynamics simulation in water resources management. *Water Resources Management* **23** (7): 1301–1323 DOI: 10.1007/s11269-008-9328-7

Wolstenholme EF. 1999. Qualitative vs quantitative modelling: the evolving balance. *Journal of the Operational Research Society* **50** (4): 422–428 DOI: 10.1057/palgrave.jors.2600700

World Energy Council. 2016. World Energy Trilemma 2016: Benchmarking the sustainability of national energy system. London. Available at: http://www.worldenergy.org/publications/2016/world-energy-trilemma-2016-defining-measures-to-accelerate-the-energy-transition/

Wu L, Elshorbagy A, Pande S, Zhuo L. 2021. Trade-offs and synergies in the water-energy-food nexus: The case of Saskatchewan, Canada. *Resources, Conservation and Recycling* **164** (June 2020): 105192 DOI: 10.1016/j.resconrec.2020.105192

Yan Q, Bi Y, Deng Y, He Z, Wu L, Van Nostrand JD, Shi Z, Li J, Wang X, Hu Z, et al. 2015. Impacts of the Three Gorges Dam on microbial structure and potential function. *Scientific Reports* **5**: 1–9 DOI: 10.1038/srep08605

Zhai Y, Zhang T, Bai Y, Ji C, Ma X, Shen X, Hong J. 2021. Energy and water footprints of cereal production in China. *Resources, Conservation and Recycling* **164** (April 2020): 105150 DOI: 10.1016/j.resconrec.2020.105150

Zhang C, Chen X, Li Y, Ding W, Fu G. 2018. Water-energy-food nexus: Concepts, questions and methodologies. *Journal of Cleaner Production* **195**: 625–639 DOI: 10.1016/j.jclepro.2018.05.194

Zhang X, Vesselinov V V. 2016. Integrated modeling approach for optimal management of water, energy and food security nexus. *Advances in Water Resources* **101**: in revision DOI: 10.1016/j.advwatres.2016.12.017

Zhao Z, Tang C, Zhang X, Skitmore M. 2016. Agglomeration and competitive position of contractors in the international construction sector. *Journal of Construction Engineering and Management*: 1–9 DOI: 10.1061/(ASCE)CO.1943-7862.0001284.

LIST OF ACRONYMS

ADB	:	Asian Development Bank
BAPPEDA	:	Local Development Planning Agency
BKP	:	National Food Security Agency
BNPB	:	Indonesian National Board for Disaster Management
BPS	:	Statistics Agency
CLD	:	Causal Loop Diagram
CP	:	Competitive Position
EBA	:	Economic Based Analysis
DLHK	:	Environmental Management Agency
FAO	:	Food and Agriculture Organization
GMB	:	Group Model Building
GRDP	:	Gross Domestic Regional Product
IRENA	:	The International Renewable Energy Agency
JR	:	Jatiluhur Reservoir
K-WEFS	:	Karawang Water-Energy-Food Security
LPDP	:	Endowment Fund for Education Agency
LP2B	:	Sustainable Food Crop Agriculture Land Policy
LQ	:	Location Quotient
MoA	:	Ministry of Agriculture
MoEF	:	Ministry of Environment and Forestry
MoEMR	:	Ministry of Energy and Mineral Resources
MoF	:	Ministry of Finance
MoPW	:	Ministry of Public Work
NGO	:	Non Government Organization
PDAM	:	Local Drinking Water Company
PERDA	:	Local Regulation
PJT II	:	Public Corporation/Reservoir Authority (Perum Jasa Tirta II)
PLN	:	State Electricity Company
RI	:	Republic of Indonesia
RPJMD	:	Regional Mid-term Planning
RPJPD	:	Regional Long-term Planning
RPJMN	:	National Mid-term Planning
RPJPN	:	National Long-term Planning
RTRW	:	Spatial Planning
SFD	:	Stock Flow Diagram
UNDP	:	United Nation Development Program
UNWWAP	:	United Nation World Water Assessment Programme
WEF	:	Water-Energy-Food

LIST OF TABLES

Table 1.1.   Summary of methodology 5

Table 2.1.   A selection of representative WEF nexus frameworks and their main
features 13

Table 2.2.   The evaluation of selected WEF nexus frameworks 16

Table 2.3.   Critiques on the concept, application, and implication of the WEF nexus 18

Table 2.4.   Summary of research related to addressing WEF nexus knowledge gaps 21

Table 2.5.   Some water, energy, and food data sources 24

Table 3.1.   Description of the three study areas 36

Table 3.2.   Average GRDP Year 2011-2015 by Industrial Origin at Constant Market
Price 2010 in West Java Province, and three study regions 37

Table 3.3.   LQ & P Values of Karawang Regency Year 2011-2015 38

Table 3.4.   LQ & P Values of Cianjur Regency Year 2011-2015 40

Table 3.5.   LQ & P Values of Bekasi City Year 2011-2015 41

Table 3.6.   Possible strategies for WEF-related sectors in Karawang Regency 48

Table 4.1.   Elements in causal loop diagrams 59

Table 4.2.   Elements of a typical GMB script 63

Table 5.1.   Initial data of K-WEFS stock flow diagrams (base year 2010) 83

Table 5.2.   Indices in K-WEFS stock-flow diagrams 84

Table 5.3.   Summary of statistical measures to test the model behaviour 85

Table 5.4.   Policy and scenario analysis 87

Table 5.5.   Model testing of selected variables (2010-2019) 91

Table 5.6.   Implication and practical recommendation on WEF-related policy and
planning 102

Table 6.1.   Scientific and societal contributions of the research 116

LIST OF FIGURES

Figure 1.1.   The importance of WEF security nexus framework 4

Figure 1.2.   Administrative and land use map of Karawang Regency, Indonesia..........7

Figure 2.1.   Schematic flow chart of the process followed in this chapter....................11

Figure 2.2.   The water, energy, and food security nexus framework (source: Hoff, 2011. Reprinted with permission) 12

Figure 2.3.   The main proposed principles and perspective for future WEF nexus concepts and frameworks 27

Figure 3.1.   Archetype competitive position chart (source: Modified from Zhao et al., 2016)...34

Figure 3.2.   Study area..36

Figure 3.3.   Competitive position of each industrial sector in Karawang Regency (2011-2015). Numbers in the bubbles refer to the sectors in Table 3.3.....39

Figure 3.4.   Competitive position of each industrial origin in Cianjur Regency (2011-2015). Numbers in the bubbles refer to the sectors in Table 3.4.....41

Figure 3.5.   Competitive position of each industrial origin in Bekasi Regency (2011-2015). Numbers in the bubbles refer to those in Table 3.5. 42

Figure 3.6.   Comparison of LQ (bars) and P Values (dots) of WEF-related sectors in the three study regions. Dashed blue line indicates LQ = 1 43

Figure 3.7.   Competitive position of WEF-related sectors based on LQ and P values..44

Figure 3.8.   Agglomeration Trends of WEF-related sectors of Karawang Regency (2000-2015)...45

Figure 3.9.   Agglomeration trends of WEF-related sub-sectors in Karawang Regency (2000-2015)...46

Figure 3.10. The trends of competitive positions of WEF sub-sectors in the years 2001 and 2015 in Karawang...46

Figure 3.11. Competitive position of WEF sub-sectors in Karawang Regency (2000-2015)...47

Figure 4.1.   The general stages in applying system dynamics GMB (a) and the process of causal loop diagram development (b) 60

Figure 4.2.   Basic concept of WEF security nexus in Karawang Regency, Indonesia..62

Figure 4.3.   The stages of GMB workshop on WEF security nexus in Karawang
               Regency 63

Figure 4.4.   Water security sub-model 66

Figure 4.5.   Energy security sub-model 67

Figure 4.6.   Food security sub-model 68

Figure 4.7.   K-WEFS nexus causal loop diagram (CLD) 70

Figure 4.8.   An example of K-WEFS model application in analysing qualitatively the
               planned interventions on water, energy, and food related sectors in the
               study area. Coloured variables, connected by thick arrows, highlight the
               feedback loops explicitly described in the text. Dark-shaded variables and
               arrows show all those variables affected taking a change in "Artificial
               ponds" as a starting point. 71

Figure 5.1.   a) Main elements of stock-flow diagram, and b) basic modes of dynamics
               behaviour 80

Figure 5.2.   Karawang Regency map 81

Figure 5.3.   K-WEFS model development stages 82

Figure 5.4.   High-level dynamics mechanism of WEF security nexus in Karawang
               Regency, Indonesia. The blue arrows represent positive causalities (i.e. a
               change in variable X causes a change in variable Y in the same direction),
               while the red arrows indicate negative causalities (i.e. a change in variable
               X causes a change in variable Y in the opposite direction) 83

Figure 5.5.   Stock-flow diagram of K-WEFS nexus model. Square boxes indicate
               stocks, thick arrows with 'clouds' indicate flows, and circles indicate
               connectors (auxiliary variables). Thin connecting arrows transmit
               information between model elements. This SFD is based on the CLD
               developed and fully described in Purwanto et al. (2019) 90

Figure 5.6.   K-WEFS nexus model validation: a) population; b) agriculture area; c)
               paddy production; and d) energy supply 91

Figure 5.7.   Model results of selected variables at base run: a) availability per person
               (APP); b) self-sufficiency level (SSL) of water, energy and food 92

Figure 5.8.   (a) water supply, water demand and imported water trends in m^3/year, (b)
               energy supply, energy demand, and imported energy trends in kWh/year,
               (c) food supply, food demand and imported non-staple food in tons/year,
               (d) environment quality index, water security index, energy security index,
               and food security index (dimensionless) 93

Figure 5.9.   Sensitivity analysis of APP water, energy, and food parameter to the positive and negative changes in the birth rate (a-c) and industrial growth: (d-f). The bold lines in graph a-f represent the baseline results. 97

Figure 5.10. The implication of scenario #1 (in-migration rate increase) on the APP and SSL of water, energy and food in the period of 2010-2030 98

Figure 5.11. The implication of scenario #2 (agricultural land conversion) on the APP and SSL of water, energy and food in the period of 2010-2030 99

Figure 5.12. The implication of scenario #3 (artificial pond and solar electricity development) on the APP and SSL of water, energy and food in the period of 2010-2030 100

Figure 5.13. The implication of scenario #4 (water, energy and food per capita consumption) on APP and SSL of water, energy and food in the period of 2010-2030...101

Figure 6.1.   The trends of gross regional domestic product (GRDP) of Karawang Regency (a) Year 2000-2010 based on constant price 2000 and (b) Year 2011-2017 based on constant price 2010 113

Figure 6.2.   The shifting of competitive position of agricultural and industrial sectors in Karawang Regency in the period of 2000-2010 and 2011-2017 113

Figure 6.3.   The proposed K-WEFS nexus framework for WEF resource evaluation and planning 117

ACKNOWLEDGEMENTS

At the end of this long journey, I would like to take a moment to acknowledge people and organizations without whom this PhD work would never have been possible. I greatly appreciate them all for all their assistance and supports.

First and foremost, I would like to express my sincere gratitude to my Promotor Prof. Charlotte de Fraiture for continuous support throughout my PhD study. You have given me the best opportunity to find my own path by reminding me of some invaluable criticisms and feedback. To my mentor, Dr. F.X. Suryadi, thank you very much for all your motivation, patience, great knowledge, and expertise. You have always been there for me, kept me in check, and over the years have become more than a teacher and partner for me. Thank you for everything. Dr. Janez Sušnik, your precious guidance and advice assisted me much in doing this research and writing of scientific publications and thesis. Thank you so much for all your time and patience in giving me correction, bringing improvement, and encouraging self-confidence to accomplish my PhD work. I feel very proud and honoured to work with you all during this PhD journey.

Thank you to all Doctoral Thesis Committees, Rector of IHE-Delft, and Rector of WUR. It goes also to the international journal reviewers to whom my paper manuscripts submitted to. I have received a lot of valuable feedback, insightful comments and encouragement, and of course challenging questions and remarks from various perspectives that have been broadened my scientific horizon. I would like to extend my gratefulness to IHE-Delft staff: Jolanda, Anique, Floor, Niamh, Ellen, Front officers, and other staff for assisting me in completing all kind of administrative things.

My sincere appreciation also goes to the Government of Indonesia in particularly the Ministry of Finance for providing scholarship through LPDP (Indonesia Endowment Fund for Education). I could not have gone through the doctoral program overseas without its financial support. I would like to express also my full gratitude to the people in The Provincial Government of West Java (Program 300 Doktor) and The Local Government of Karawang Regency for all the supports during my PhD work. Without their precious supports, it would not be possible to carry out this research.

Some special words of gratitude go to my PhD colleagues with whom I have shared both anxiety and excitement: Aklan, Eiman, Alex, Tarn, Adey, Annelieke, Hieu, Maria, Mohannad, A Mahmoud, A Elghandour, Thaha, Polpat, Ha, Meseret, and other colleagues for the warm and stimulating discussions, 'good morning and see you', laughs, smiles, and also serious faces during the deadlines. Thank you, buddies! For making my PhD days with many amazing moments we had. To all my Indonesian friends at IHE-Delft, Pak Dikman, Sebrian, Mba Wiwit, Mba Fiona, Mba Yuli, Mas Sonny, Mba Mitha, Teh Selvi and also all PhD and Master students at IHE, TU Delft, and WUR that I could not mention here. Thank you so much for keeping me on the right paths through your scientific and personal suggestions. For the big family of Keluarga Muslim Delft (KMD) and MSA-IHE, I will never forget the warm and friendly atmosphere of the entire activities. Thank you for the real brotherhood and great supports.

I would like also to express my deepest appreciation to Bupati Karawang, dr. Cellica Nurrachadiana for allowing me to pursue my PhD in the Netherlands. I also sincerely acknowledge my extraordinary people in Karawang Regency, Head and Staff of DLHK, BKPSDM, DPMPTSP, DLHK, BAPPEDA, and other institutions, in particularly participants in Group Model Building (GMB) workshop. Thank you for all your great support, resources, data, and collaboration. I will always remember that indispensable occasions in my life.

Last but certainly not least, I would like to convey my sincerest gratitude to my big family in Indonesia, Ayah & Ibu, Om Bing & Tante Tuti, my brothers and sisters, and in particularly my lovely parents for their sacrifices, spiritual supports, and encouragement. I am truly indebted to both of them, and I hope this is the last time I stay far away from them in their old age moment. To my late mother, my late supervisors (Prof. Robiyanto and Prof. Iwan K), may you all be granted the highest place in heaven. To my beloved little family, my beautiful wife Ana Natalina, and my great sons and daughter: Abyan, Aftah, and Alifa, thank you for delivering your love, patience, and smiles. I am not a perfect husband and dad, but I promised to be much better in loving you all until the end of my life. You are all like the candles in the dark for me. Please accept my deepest apologies.

To all of you I have not mentioned personally on this limited page, my greatest gratitude and acknowledgment for all of you are beyond the words I have. You should know that all your contributions and supports during my PhD work were worthwhile for my entire life. Thank you so much.

Delft, the Netherlands,

Aries Purwanto

ABOUT THE AUTHOR

Aries Purwanto was born in Karawang, Indonesia. He received his Bachelor Degree in Mining Engineering in 2003 from University of Sriwijaya, Indonesia. Afterwards, he was awarded a scholarship from BAPPENAS and STUNED to continue his master study in double-degree programme on Integrated Lowland Development and Management Planning in University of Sriwijaya & UNESCO-IHE, Delft and graduated by 2013. He has been working as a Development Planner in Karawang Regency,  Indonesia for more than 10 years. Started from September 2016, Indonesia Endowment Fund for Education (LPDP), Ministry of Finance of Indonesia provided him a full financial support to undertake Ph.D. research in Land and water Management at IHE-Delft and Wageningen University and Research (WUR) the Netherlands as Partner University. His research explores the water, energy, and food (WEF) security nexus in the local context. The findings are expected to assist local government and stakeholders in evaluating and planning the complex WEF security system in their local regions and bring positive impacts to the national WEF security targets. During his PhD work, he has participated in several courses, seminars and scientific workshops. Furthermore, he has delivered his PhD research results through several national and international conferences, and has published several scientific papers both in the national and international peer-reviewed journals. He finally received the SENSE diploma by June 15, 2021.

Journal publications

1. **Purwanto A.**, (2017), *Economic base analysis of water, energy, and food related sectors: a case of West Java Province*, Creative Research Journal, ISSN: 2579-9231, Vol. 03 No. 02 December, 73–90, DOI: http://dx.doi.org/10.34147/crj.v3i02.101 (Published).

2. **Purwanto A.**, Sušnik J, Suryadi F.X., de Fraiture, C (2018), *Determining strategies for water, energy, and food-related sectors in local economic development*, Sustainable Production and Consumption 16, 162–175 Elsevier, DOI: https://doi.org/10.1016/j.spc.2018.08.005 (Published).

3. **Purwanto A.**, Sušnik J, Suryadi F.X., de Fraiture, C (2019), *Using group model building to develop causal loop mapping of the water-energy-food security nexus in Karawang Regency, Indonesia*, Journal of Cleaner Production (JCLP), Elsevier 240, DOI: https ://doi.org/10.1016/j.jclepro.2019.118170 (Published).

4. **Purwanto A.**, Sušnik J., Suryadi F.X., de Fraiture C., (2020), *Quantitative simulation of the water-energy-food (WEF) security nexus in a local planning context in Indonesia*, Sustainable Production and Consumption Journal, Elsevier, DOI: https://doi.org/10.1016/j.spc.2020.08.009 (Published).

5. **Purwanto A.**, Sušnik J., Suryadi F.X., de Fraiture C., (2021), *Water-energy-food nexus: Critical review, practical applications, and prospects for future research*, Sustainability 13, 1919, MDPI, DOI: https://doi.org/10.3390/su13041919 (Published).

6. D. Daniel, Prawira J., T. P. Al Djono, S., Subandriyo, Rezagama A., **Purwanto A.**, (2021), *A System Dynamics Model of the Community-based Rural Drinking Water Supply Program (PAMSIMAS) in Indonesia*, Water 13, 507, MDPI, DOI: https://doi.org/10.3390/w13040507 (Published).

Conference paper & proceeding

1. **Purwanto A.**, Sušnik J, Suryadi F.X., de Fraiture, C (2019), *Causal loop diagram of WEF security nexus: An implementation of group model building approach*, ICID Conference Proceeding, 3rd World Irrigation Forum, Bali-Indonesia, 1-7 September 2019.

2. **Purwanto A.**, (2019), *Can agriculture and industry be synergized in local economic development?*, International Conference on Indonesian Development, 19-21 September, Rotterdam, the Netherlands.

Netherlands Research School for the
Socio-Economic and Natural Sciences of the Environment

D I P L O M A

for specialised PhD training

The Netherlands research school for the
Socio-Economic and Natural Sciences of the Environment
(SENSE) declares that

Aries Purwanto

born on 27 March 1980 in Karawang, Indonesia

has successfully fulfilled all requirements of the
educational PhD programme of SENSE.

Wageningen, 15 June 2021

The Chairman of the SENSE board

Prof. dr. Martin Wassen

the SENSE Director of Education

Dr. Ad van Dommelen

The SENSE Research School has been accredited by the Royal Netherlands Academy of Arts and Sciences (KNAW)

K O N I N K L I J K E   N E D E R L A N D S E
A K A D E M I E   V A N   W E T E N S C H A P P E N

The SENSE Research School declares that Aries Purwanto has successfully fulfilled all requirements of the educational PhD programme of SENSE with a work load of 32.6 EC, including the following activities:

SENSE PhD Courses

o   Environmental research in context (2017)
o   Research in context activity: 'Initiating and organizing System Dynamics Group Model Building (GMB) of Water-Energy-Food Security Nexus with stakeholder workshop in Karawang Regency, Indonesia 17 October 2018'

Other PhD and Advanced MSc Courses

o   Scopus advanced tips and tricks: a changing world of research, IHE Delft (2016)
o   EndNote training session, IHE Delft (2016)
o   Open source software for pre-processing GIS data for hydrological models, IHE Delft (2017)
o   Academic writing course, IHE Delft (2018)
o   Paper writing course, IHE Delft (2018)

Management and Didactic Skills Training

o   Supervising two MSc students with thesis entitled 'Analysis of watershed erosion management: Analysis of watershed erosion management: A case study of Rodrigues Island (2018) and 'Spatial water, energy, and food (WEF) nexus balance in local region context: A case of Karawang Regency, Indonesia (2019)
o   Teaching in Tailor Made Training modules 'Watershed delineation and soil erosion simulation using QGIS' (2017), 'Group model building on WEF security nexus (2019) and ' Modernized irrigation and water accounting for irrigation' (2019)

Oral Presentations

o   *Causal loop diagram of WEF security nexus: An implementation of group model building approach.* 3rd International Conference of World Irrigation Forum 2-5 September 2019, Bali, Indonesia
o   *Can agriculture and industry be synergized in local economic development?* 2nd International Conference on Indonesia Development, 19 September 2019, Rotterdam, The Netherlands

SENSE coordinator PhD education

Dr. ir. Peter Vermeulen

T - #0422 - 101024 - C174 - 240/170/10 - PB - 9781032076454 - Gloss Lamination